Proceedings of the Regional Workshop

Wastewater Treatment and Reuse for Metropolitan Regions and Small Cities in Developing Countries

Editor: Edmilson Santos de Lima

September 11-17, 2016 – Recife, Brazil

Proceedings of the Regional Workshop

Wastewater Treatment and Reuse for Metropolitan Regions and Small Cities in Developing Countries

Editor: Edmilson Santos de Lima

September 11-17, 2016 – Recife, Brazil

Issue Editor

Prof. Dr. Edmilson Santos de Lima

Federal University of Pernambuco (UFPE), Center of Technology and Geosciences, Recife, Pernambuco, Brazil

delima@ufpe.br

Exceed Chairman & Editor-in-Chief

Prof. Dr.-Ing. Norbert Dichtl

Technische Universität Braunschweig, Institute of Sanitary and Environmental Engineering, Braunschweig, Germany

n.dichtl@tu-bs.de

Publishing Editor

Prof. em. Dr. mult. Dr. h.c. Müfit Bahadir

Technische Universität Braunschweig, Leichtweiss Institute, Exceed, Braunschweig, Germany

m.bahadir@tu-bs.de

This publication was financed by the German Academic Exchange Service (DAAD) and the Federal Ministry for Economic Cooperation and Development (BMZ).

Bibliografische Information der Deutschen Nationalbibliothek
Die Deutsche Nationalbibliothek verzeichnet diese Publikation in der
Deutschen Nationalbibliografie; detaillierte bibliografische Daten sind im Internet
über http://dnb.d-nb.de abrufbar.
1. Aufl. - Göttingen: Cuvillier, 2016

© CUVILLIER VERLAG, Göttingen 2016
  Nonnenstieg 8, 37075 Göttingen
  Telefon: 0551-54724-0
  Telefax: 0551-54724-21
  www.cuvillier.de

1. Auflage, 2016
Gedruckt auf umweltfreundlichem, säurefreiem Papier aus nachhaltiger Forstwirtschaft.

  ISBN 978-3-7369-9454-6
  eISBN 978-3-7369-8454-7

CONTENT

PREFACE

This proceedings book brings 14 papers selected out of 27 presented at the International Expert Workshop on *"Wastewater Treatment and Reuse in Metropolitan Regions and Small Cities in Developing Countries"*, held in Recife, Brazil September 14–16, 2016 within the framework of the EXCEED SWINDON Project. Participants were from 9 different countries (Argentina, Brazil, Cameroon, Colombia, Germany, Jordan, Mexico, Thailand and Turkey) that made the workshop very broad in participation and expertise.

The main objectives of the workshop were to share experiences, knowledge, and research outcomes between the different participants, and to discuss the main problems and challenges in the field of wastewater treatment and reuse for Metropolitan Regions and Small Cities in Developing Countries. In addition, the workshop dealt with suitable technologies for domestic sewage treatment, removal of recalcitrant substances and micro-pollutants from domestic sewage and industrial wastewater, and reuse of treated wastewater in urban areas, industry and agriculture. Another important goal of the workshop was to further strengthen the expert network within the EXCEED SWINDON Project.

Two independent experts, Prof. Dr. Lourdinha Florencio and Prof. Dr. Mario T. Kato reviewed all papers submitted to the workshop in terms of their scientific content, and the Publishing Editor of the Proceedings Book, Prof. Dr. Müfit Bahadir made the final publishing review. As the Issue Editor of these proceedings I cordially acknowledge their invaluable contributions while publishing this book.

I hope and am confident that the readers will gain benefit from reading these papers.

Prof. Dr. Edmilson S. de Lima
Centro de Tecnologia e Geociências, Universidade Federal de Pernambuco, Brazil

NATURAL WASTEWATER TREATMENT AND WATER QUALITY MONITORING SYSTEM FOR EFFECTIVE REUSE

Sergio Eduardo Abbenante, Martín Damián Vergara

Universidad Tecnológica Nacional – La Plata – Buenos Aires – Argentina
(ing.abbenante@gmail.com)

Keywords: Argentina, Phytoremediation, Wastewater treatment, Wetlands, Software monitoring

Abstract

There is a large quantity of wastewater in Argentina, of which only about 10% is treated to reuse. The rest is sent directly to the rivers without any treatment, which increase their pollution. It is estimated that 73% of water in rural areas of Argentina is used for agriculture and livestock. An investigation concluded that there are few efficient methods in rural areas of Argentina to mitigate this problem. The solutions that have been proposed are too expensive to be applied in these areas. For this reason, the works are not completed because of high cost of investment and maintenance. From these findings, an easy handling and low cost invest-ment tool of wastewater treatment as well as a system designed with "free software" tools to control and to monitor its quality was developed as a solution of this problem and to extend it elsewhere in the world with similar needs. First, the characteristics of design and implemen-tation of wastewater treatment plant in rural areas through wetlands using native plants are presented in this study, and the details of this technique are explained. Then a solution to monitor and to optimize the water quality through software for their effective reuse is presented. Finally, the application of the two techniques and the performance, when both methods are applied, are presented in a scenario. This study has shown that by applying these techniques the treatment area will gain benefits from using low cost implementation tools, easy handling techniques and optimization of processes.

1 Introduction

There is a large quantity of wastewater in Argentina and only about 10% of this amount is treated to reuse. The rest is sent directly to the rivers without any treatment that increase their pollution. The Matanza-Riachuelo River (MR), a tributary of the Río de la Plata (La Plata River), is the most contaminated river basin in Argentina and is considered one of the most polluted water bodies in the world [1]. Pollution levels in Buenos Aires' rivers are so high that they have been considered "open sewers", making pollution the greatest environmental risk for the metropolitan area. Pollution levels have increased steadily as urbanization and indus-trial growth have continuously increased as well in the metropolitan and rural areas of Buenos Aires. It is estimated that more than 4,000 industrial facilities are located in the lower and middle sections of the basin. Almost all of these industries discharge untreated effluents into the drainage system or directly into the river. In addition to high levels of organic pollution, these discharges contain toxic contaminants such as heavy metals from petrochemical industries, tanneries, and meat processing facilities [2]. More specifically, the Riachuelo River has levels of lead, zinc and chromium 50 times higher than the legal limits in Argentina; 25% is from industrial sewage and waste, while the remaining 75% originates from domestic sources [3].

The sudden onset of environmental and social degradation of the basin has resulted from limited investment in public infrastructure, poor environmental management, lack of adequate urban and industrial planning, and limited public infrastructure investment. Also, in areas of intensive farming, uncontrolled use of agrochemicals contributes to pollution of water resources [4].It is estimated that 73% of water in Argentina is used for agriculture and livestock [5]

About 65% of the population has access to sewer system, but in rural areas only 1% of the population has the same. In rural areas, wastewater that comes from human use, industries, agriculture and livestock is not treated properly. In these areas, it is necessary to apply a low-cost method for the treatment of wastewater taking into account the low implementation costs, easy handling and low cost maintenance.

2 Material and Methods

2.1 Design and implementation of wastewater treatment plant through wetlands
Wetlands are one of natural systems that can be used for wastewater treatment and pollution control [6].

2.1.1 Type of wetland system used - Subsurface Flow (SSF)
In subsurface-flow systems, the effluent may move either horizontally, parallel to the surface, or vertically from the planted layer down through the substrate and out. This system has several advantages but also some disadvantages, as shown below:

Advantages of this technique are
- Operational simplicity, limited to the removal of waste pre-treatment, and cutting and removal of vegetation, when it is dried,
- Flexible system and low sensitivity to changes in flow and load,
- The vegetal biomass of the sediment acts as insulation, ensuring microbial activity throughout the year,
- Seamless integration into rural areas; no environmental impacts sound, no odor generated,
- Potential use of plant biomass produced,

- Subsurface horizontal-flow wetlands are less hospitable to mosquitoes, as there is no water exposed to the surface, whose populations can be a problem in surface-flow constructed wetlands, and
- Subsurface-flow systems require less land area for water treatment.

Disadvantages of this technique are
- Larger land area is required for implementation than conventional technologies of treatment, and
- In the wetlands of free flow, as water flows over the substrate surface, it can cause the proliferation of mosquitoes.

Characteristics of Subsurface Flow (SSF) wetlands are (i) one or more cells (usually two in number), (ii) ground limited by partitions (landfills), (iii) waterproofed with polyethylene film, (iv) filled with porous material (gravel, stone, sand or soil), (v) planted with emergent aquatic plants, (vi) bed depth of 0.7 - 1.2 m and water depth 0.6 - 0.7 m, (vii) their fund is constructed with slope of 1-2%, (viii) perforated pipe entrance used surface perpendicular to the flow, (ix) output per pipe drilled perpendicular to the flow line, (x) hydraulic loadings applied between 2 and 20 cm/d, and (xi) land areas between 0.5 and 5 ha per 1000 m^3/d) are required.

2.1.2 Phytoremediation

For this system, the technique of phytoremediation is used [7]. It is a cost-effective plant-based approach of remediation that takes advantage of the ability of plants to concentrate elements and chemical compounds from the environment and to metabolize various molecules in their tissues. It refers to the natural ability of certain plants called hyperaccumulators to bioaccumulate, to degrade, or to render harmless contaminants in soils, water, or air. Toxic heavy metals and organic pollutants are the major targets for phytoremediation. The figure 1 describes the phytoremediation process [8]. The most important characteristics of this method are the following:

Phytoextraction – uptake and concentration of substances from the environment into the plant biomass,

Phytostabilization – reducing the mobility of substances in the environment, for example, by limiting the leaching of substances from the soil,

Phytotransformation – chemical modification of pollutants as a direct result of plant metabolism, often resulting in their inactivation, degradation (phytodegradation), or immobilization (phytostabilization),

- *Phytostimulation* – enhancement of soil microbial activity for the degradation of pollutants, typically by rhizosphere organisms that associate with roots; this process is also known as rhizosphere degradation. Phytostimulation can also involve aquatic plants supporting active populations of microbial degraders, as in the stimulation of atrazine degradation by hornwort.

- *Phytovolatilization* – removal of substances from soil or water with release into the air, sometimes as a result of phytotransformation to more volatile and/or less polluting substances, and

- *Biological hydraulic containment* – some plants, like poplars, draw water upwards through the soil into the roots and the plant biomass. This decreases the transport of soluble pollutants downwards, deeper into the soil and the groundwater.

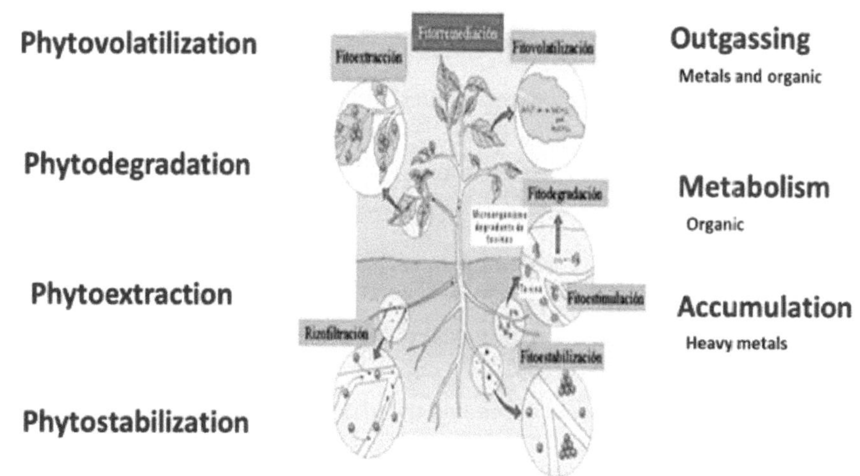

Figure 1: Phytoremediation process

2.2 Monitoring and optimization of water quality through software for their effective reuse
The software will act as decision making tool. Using specific informatics tools to import data from different sources and importing manual data on it, one can obtain important information to define the destination of water. Once the user inserts the data into the software, the informatics tool will show the results processing the inputs. It is a very useful tool to decide the correct destination of the treated water. This software was developed using open source tools [9-11].

2.2.1 Sample information
To evaluate the water quality in each water income, the software makes use of sample information to decide on outgoing water. The method taking samples is divided in this project in two levels:

Sampling Level 1 - Minimal equipment inversion and quick evaluation kits; this method can be learned with basic training.

Sampling Level 2 - Inversion in precision machines, more training, low number of participants

2.2.2 Web-based system software for information management
The web-based system software was developed using free license technologies to web develo-ping: PHP 5, Apache web server, MySQL databases [10-12]. The information and services must be accessible to all and able to be used with all navigation devices. Furthermore, it must have clear and simple contents as well as simple navigation mechanisms. For information safety, backups, passwords, digital certificates, cryptography and remote backup systems are very important aspects of data security. It is very important to have concurrent access to the system from all navigation devices at the same time at different geographic points.

2.2.3 Development software advantages
It is very important to use easy handling techniques and friendly screens. Quick access to infor-mation through listings and graphs is required. Implementation costs are low compared with sophisticated systems. It is free license software.

2.2.4 Key Features
About the data obtained, decisions can be taken regarding the destination of treated water in terms of

- Access to information from any geographical point (working area, university),
- Evolution of different parameters,
- Analysis of requirements to avoid exceeding the allowable limits of pollutants,
- Analysis of indicators, and
- Information for decision-making (water recirculation, use for irrigation).

3 Results and Discussion

3.1 System scheme
This application is an alternative for wastewater treatment. It is developed to use in rural areas of Argentina, where shortage of water for irrigation and livestock causes a problem, and the wastewater is currently discharged untreated into rivers. The overall process scheme is shown in figure 2.

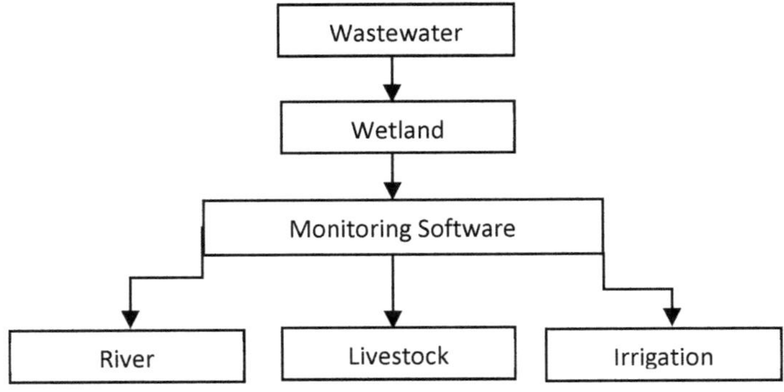

Figure 2: Scheme of full system steps

3.2 System process

Wastewater is sent first to the pre-treatment plant. After that it goes directly to the pools, where the selected plants act using a pre-established configuration. After this process, water is prepared to go to different destination: river, irrigation or to pretreatment to use it for livestock. The configuration can be modified due to the quality of water, if it is necessary to be improved (figure 3).

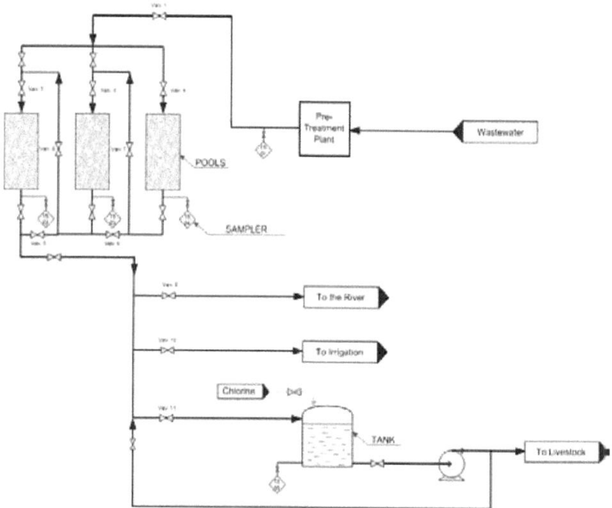

Figure 3: Scheme of water process

3.3 Wetlands monitoring diagram

The user inserts each the data of sampling into the system. Thereafter, it will be possible to answer whether the water is suitable to be used for different purposes. All data, which are automatically obtained, will be imported directly into the system (figure 4).

The software has a several tools. For example, it will also evaluate the quality of water if one parameter is out of range.

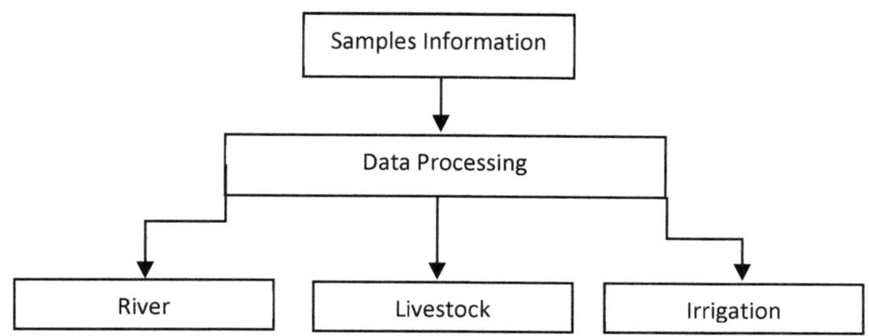

Figure 4: Scheme of software tool

4 Conclusions

The system developed and described in this paper shows low implementation costs, easy handling, natural treatment system using native plants and easy access for the optimization of the techniques applied. With this, it is possible to increase water reuse for livestock and irrigation taking into account the allowable limits of pollutants. Dynamic tools to improve and to optimize the process with the development of web-friendly specialized software accessible from any device could be realized and also applied elsewhere under similar conditions. So it is an opportunity to improve the effluent quality of wastewater and to minimize the impacts on water bodies.

5 Acknowledgements

The authors wish to thank to DAAD and Exceed Swindon project that made the participation at the workshop possible. We were also glad to share our knowledge with participants of future events and projects.

6 References

[1]   https://fororiodelaplata.wordpress.com/2009/07/29/el-rio-de-la-plata-el-tercer-rio-mas-contaminado-del-mundo/

[2]   http://ri.conicet.gov.ar/handle/11336/1368

[3]   http://www.winisisonline.com.ar/tea/info/BD00183.PDF

[4]   http://edant.clarin.com/diario/2005/01/22/opinion/o-03602.htm

[5]   http://www.cepal.org/drni/proyectos/samtac/inar00200.pdf

[6]   U.S. EPA (1983). https://www.epa.gov/wetlands

[7]   United Nations Environment Programme. Newsletter and Technical Publications Phytoremediation: An Environmentally Sound Technology for Pollution Prevention, Control and Remediation (1994) http://www.unep.or.jp/Ietc/Publications/Freshwater/FMS2/2.asp

[8]   http://fourthcornernurseries.com/the-use-of-aquatic-plants-to-treat-waste-water/

[9]   «Proyecto GNU - Licencias de software libre incompatibles con la GPL», September 1983 http://www.gnu.org/licenses/license-list.html#GPLIncompatibleLicenses

[10]   Phpnet. https://secure.php.net/archive/2016.php#id2016-08-18-1

[11]   MySql database. https://dev.mysql.com/doc/relnotes/mysql/5.7/en/

[12]   Apache server. "About the Apache HTTP Server Project". Apache Software Foundation. Archived from the original on 7 June 2008. Retrieved 2008-06-25. http://httpd.apache.org/ABOUT_APACHE.html

SELECTION OF PLANT SPECIES USED IN WASTEWATER TREATMENT

[1]Eduardo Saldanha Vogelmann, [2]Gabriel Oladele Awe, [3]Juliana Prevedello

[1] Institute of Biological Sciences, Federal University of Rio Grande, Rio Grande, Brazil (eduardovogelmann@furg.br).

[2] Department of Soil Resources and Environmental Management Department, State University, Ado Ekiti, Nigeria.

[3] Institute of Oceanography, Federal University of Rio Grande, Rio Grande, Brazil.

Keywords: biological treatment; constructed wetlands; plant species; water quality; waste treatment.

Abstract

The treatment of effluents using biological methods such as plants is a complex waste management option comprising water, substrate, plant roots, and a large number of microorganisms which interrelate. A major advantage of this system is that it can be implemented *in situ* where the effluent is produced, with low cost of operation, low energy consumption and operational simplicity. The treatment is basically using plants to utilize nutrients contained in the effluent and convert to green mass, in other words the plants acting as extractors of macro- and micro-nutrients in the effluent material. In addition, such plants may also extract or permit the possibility of transforming materials containing heavy metals and toxic organic compounds that may appear difficult to treat. However, in order to succeed in such treatment option, a selection of plants should first be carried out, which is based on some criteria such as: (i) good natural adaptation to the local climate; (ii) rapid growth and high biomass production; (iii) nutrient absorption capacity; (iv) adaptation and ease of propagation; (v) good root development; (vi) oxygen transfer capacity to the roots by creating aerobic environment. However, due to the great diversity of flora, further research is needed in relation to the evaluation and selection of plant species having potentials for use in wastewater treatment in constructed wetlands.

1 Introduction

Over the years, the number of treatment plants for both domestic and industrial wastewaters in Brazil has increased tremendously, especially in urban areas, due to increased environmental awareness by the society, improved supervision, enactment and enforcement of strict environmental laws by government [4-10]. The increase in use of treatment systems has helped to meet the search for minimizing environmental impacts caused by the release of effluents, especially in water bodies. As one of the requirements of sewage and wastewater treatment systems is to reduce potential environmental contamination, the one that returns back to the environment should be an effluent that is safe in accordance with standards set by environmental legislation [11-18].

Even with a number of advantages and benefits provided by the various wastewater treatment systems, one of the main obstacles that have limited the implementation of these systems in the past decades is the high initial capital required for the construction of a plant as well as skilled staff needed to operate the systems, which even become more technical each day [9]. In addition, in rural areas where the population density is low, the establishment of modern sewage treatment plants is not always economically feasible, especially in the case of wastewater treatment, simply because there are no collection networks which could be used to centralize the collection of the effluents [20]. This difficulty shows the need for an alternate and affordable option that should be urgently considered and implemented with a view to obtaining low-cost technology, but which should give efficiency that is equal or even greater than that of conventional treatment plants used in the treatment of domestic and industrial effluents.

This need for treatment method of high efficiency, simplicity of operation and associated low investment has driven the use of alternative wastewater treatment systems that simulate the phenomena that occur spontaneously in nature, such as those seen in marshy areas and wetlands, where microorganisms and plant species can naturally purify the water [1-20]. This principle of wastewater treatment technique using aquatic plants, also called constructed wetlands or root zone, has become a well-recognized and recommended option, mainly because of its low cost, high effectiveness in reducing organic matter, assimilation of nutrients by plants and containment or elimination of toxic substances in effluents [23].

Wastewater treatment using plants, though it appears simple, is a complex management comprising effluent substrate, plant root and a wide range of microorganisms that interact to improve water quality [1-17]. In this system, the effluent that passes through the substrate undergoes a process of purification by means of certain processes including filtration and chemical precipitation by water contact with the supporting medium, retention of suspended particulate matter, chemical transformations, predation and natural reduction of pathogenic organisms, besides plant activities of growth and removal of nutrients by the root system thereby improving the physical-chemical conditions of the substrate [20]. Thus, the use of plants for sewage treatment has recently become an emerging alternate technology that is very effective than conventional systems [18]. Another advantage is that this system can be implemented *in situ* where the wastewater is produced (in the form of decentralized system), besides it can be operated by people of low educational level, and have low energy requirements [9]. In addition to high efficiency and low cost of investment, highlighted other advantages including the improvement of environmental quality, the landscape effect, the creation and restoration of ecological niches and large production of biomass, which can then be used as animal feed or for energy production, thus helping in the perspective of income generation [23].

Due to the numerous advantages presented by wastewater treatment systems using plants, a new research field has evolved for better understanding of the processes underlying the inter-relations between the different species of plants and microorganisms that optimize the use of these systems for different types of contaminants in domestic and industrial effluents [11-22]. However, one of the critical points in the design of this system also lies in the choice of plant species along with other variables of fundamental importance for the successful treatment of effluents in constructed wetlands [9].

The selection of plant species to be used is crucial to the success of the use of this system, because in addition to the microorganisms, plants are responsible for extracting nutrients from wastewater for growth, thus acting as extractors of macro and micronutrients during the effluent treatment [5-10]. Plants can still extract or permit the transformation of substances containing heavy metals and toxic organic compounds difficult to be treated by conventional methods of wastewater treatment [7-11]. This occurs mainly during the growing season, when plants uptake macronutrients, nitrogen (N) and phosphorus (P) as well as micronutrients (including metals), which are accumulated in leaves or translocated to accumulating organ, such as roots and rhizomes [5].

However, for the smooth running of effluent treatment process in constructed wetlands, it is essential to select the most suitable species. For this, it is essential to know in advance the particular kind of behavior in different climatic conditions and especially the adaptability and tolerance of particular chemical composition of the effluent to be treated [20]. In this regard, the need for new studies were reiterated that seek to determine the capabilities of different plant species and to define the potential use of each in the treatment of wastewater of different characteristics with a view to adjusting and enhancing the action of plants and bacteria in constructed wetlands, because with this, the effectiveness of treatment will be significantly increased, and enable the release of a good quality effluent to the environment.

2 Use of plants in constructed wetlands systems

For the growth and development of plants, the absorption of water and nutrients is necessary, and the uptake of these nutrients is carried out mainly by the roots [11-22]. It is exactly this process that is the guiding principle in the use of plants for the purification of contaminated water, because, depending on the plant species, a high amount of these nutrients are assimilated and converted into biomass, getting accumulated and temporarily immobilized in the plant tissue, which is harvested and removed from the site [3]. If there is no harvest, the nutrients that were incorporated into the plant return to the water by decomposition processes [21]. Apart from being nutrients, elements such as phosphorus and nitrogen are problematic when found in high concentrations in water bodies, therefore, their replacement in crops is necessary, mainly due to large extraction of these elements by grains and fruits, which are frequently harvested [22]. Thus, plants used in wastewater treatment can become interesting sources of nutrients and micronutrients required to maintain the fertility of agricultural soils. This proposition was further supported by [13] who evaluated the potentials of water hyacinth (*Eichhornia crassipes*) in nutrient uptake and concluded that in addition to being a plant capable of removing good amounts of nutrients such as phosphorus and nitrogen, it showed great potentials of accumulating zinc and chromium (for example, accumulated Zn and Cr contents were 3.542 and 2.412 mg g^{-1} dry matter, respectively), as well as successfully removed 84% and 94% of Cr and Zn, respectively, when in direct contact with concentrated solution of these metals.

Another relevant aspect and also considered in plant selection process is to analyze the effluent characteristics to be treated, or in other words, to identify the main constituents of the effluent, the removal mechanisms, or physical, chemical and biological processes that should be involved (Table 1). This should be set beforehand so that plants possessing better development and adaption in certain condition could be defined. A good example is nitrogen, which is needed abundantly by plants, especially grasses, but when associated with organic particles, it is strongly adsorbed in organic matter and not readily available to plants. In this case, filtration is needed followed by sedimentation of the particles by the root system in association with nitrification and denitrification activities of microorganisms, which assist in the removal of nitrogen from effluents [5-20]. Thus, it is evident the primordially use of plants by nitrogen, preferably without the ability to fix this element symbiotically (when effluent is considered the sole source of nitrogen), with abundant root system, make it possible to exploit higher volume of flooded area beside contributing as filtering or screen agent, facilitating the sedimentation of solid particles.

In the tropics, there are wide species of vegetation with potentials in the treatment of effluents, with many of these species possessing good tolerance to flooding or flooded sites [4]. Fundamentally, any vegetation selected for the treatment of effluents should be the one that tolerates permanently saturated areas or submerged with constant flow of contaminants from various types and concentrations [21]. Furthermore, native species should be preferred due to their ease of adaptation and grow in prevailing weather conditions [21]. Exotic species can be used only if they have been introduced to the region or when they did not show any invasive characteristics, with a view to preventing them from becoming a plague in the event of an escape from the treatment system [22].

It is also important to note the disposition of plant tissues in constructed wetlands because some species require special features during construction, especially in relation to fixing the substrate, while others do not tolerate high rate of effluent loading because of not having fixing structures in the substrate, crowding of the tank corners could occur [2-21]. In this regard, despite being generically referred to as aquatic plants, some important differences must be observed regarding the choice of plants to be used, mainly because it is desirable to ensure optimal conditions for development so that it can express its potential growth and contribute more effectively in the treatment of effluents [22].

Table 1: Principal mechanisms of pollutant removal

Pollutant	Process or mechanism of removal
Total soluble solids	Sedimentation and/or filtration
BOD_5	aerobic or anaerobic degradation process Sedimentation of organic particles
Nitrogen	Nitrification and denitrification Process of volatilization of ammonia Nitrogen uptake by roots
Phosphorus	Reactions of adsorption and precipitation with some cations Phosphorus absorption by the roots
Pathogenic organisms	Sedimentation and filtration or UV rays action Excretion of antibiotics by plants and other bacteria

Source: Adapted from [19].

In this context, it could be basically differentiated in seven major biological forms (Figure 1). However, the use of epiphytes (live on other plants) and amphibian species in constructed wetlands is not generally promising. In the case of epiphytes, many of them require a pre-existing vegetation for their installation, and therefore, may affect the development of the plant used as support and at the same time do not show great potential in the absorption of effluent nutrients, since part of the nutrients can be absorbed from the air or remnants accumulated on the supporting vegetation [22]. For the amphibious plants, in most cases they do not support permanently flooded environment and may in some cases be used only on slopes or river banks, where the contact of the roots with the effluent is reduced [4]. Below is list of the major groups of aquatic plants, as well as the biotope, showing the order of interest and applicability in constructed wetland systems:

1. *Emerging plants*: plants rooted in the sediment with leaves staying above the water (Fig. 1). Ex.: *Juncus effusus, Scirpus giganteus, Sangitary lancifolia* and *Typha domingensis*.

2. *Fixed floating plants*: plants rooted in the sediment with leaves floating on the water surface. Ex.: *Victoria Amazonica* and *Nymphaea lotus*.

3. *Free floating plants*: plants freely floating on the water surface. Usually its maximum development occurs in locations protected by the wind. Ex.: *Azolla Anabaena, Lemna valdiviana, Pistia stratiotes* and *Spirodela polyrhiza*.

4. *Fixed submerged plants*: plants rooted in the sediment growing fully submerged in the water. They can grow to great heights, depending on the availability of light. Most have their reproductive organs floating on the surface. Ex.: *Elodea canadenses, Mayaca fluviatilis, Potamogeton perfoliatus* e *Egeria densa*.

5. *Free submerged plants*: plants which have rhizoids undeveloped and remain floating submerged in water in areas of low turbulence. They are usually attached to the petioles and stems of aquatic plants of floating leaf. Ex.: *Ceratophyllum demersume, Utricularia graminifolia*.

Figure 1: Aquatic macrophytes scheme adapted from [15]. (1) amphibian; (2) emerging; (3) fixed floating; (4) free floating; (5) fixed submerged; (6) free submerged; (7) epiphyte

3 Selection of species used in constructed wetlands systems

In recent years, prominence has been given to rooted aquatic plants, mainly due to their greater root surface and anchoring in the tank as well as not suffering from direct influence of effluent flow [21]. However, the selection of plants should not be so simplified and other characteristics that must also be observed are:

1. Ease in getting seedlings, seeds or vegetative propagules;
2. Good natural adaptation to the local climate during all seasons of the year;
3. Regeneration and ease of natural propagation;
4. Resistance to the occurrence of pests and diseases;
5. Rapid growth and high biomass production during all seasons of the year;
6. Good root development with ease of anchoring to the substrate;

7. High capacity of absorption of nutrients;

8. Nutrient storage capacity for accumulation of bodies;

9. Oxygen transfer capacity to the roots creating an aerobic environment.

For these characteristics, studies have been conducted in different countries in search for selection of plants for use in constructed wetlands. Among the various options analyzed, some have been gaining momentum, mainly because of the good adaptation to adverse conditions such as low and high temperatures, high efficiency in the removal of nutrients and heavy metals from the wastewater. In addition, features such as fast growing and high potentials for extraction and conversion of nutrients into biomass are indispensable; some prominent options are listed in Table 2.

Table 2: Some aquatic plants used in constructed wetlands

Family	Scientific name	Common name	Reference
Alismataceae	*Sagitaria lancifolia*	Bulltongue arrowhead	[14]
Araceae	*Pistia stratiotes*	Water lettuce	[10-11]
Araceae	*Lemna valdiviana*	Duckweed	[2]
Araceae	*Spirodela sp.*	Duckweed	[2]
Ceratopphyllaceae	*Ceratophyllum demersum*	Hornwort	[7]
Cyperaceae	*Scirpus californicus*	Bulrushes	[5]
Cyperaceae	*Eleocharis sphacelata*	Spike sedge	[3]
Cyperaceae	*Cyperus giganteus*	Umbrella plant	[14-16-17]
Cyperaceae	*Cyperus involucratus*	Umbrella plant	[11-16]
Cyperaceae	*Scirpus validus*	Soft stem bulrush	[6-8]
Hydrocharitaceae	*Elodea canadensis*	Waterweed	[12]
Hydrocharitaceae	*Egeria densa*	Waterweed	[7]
Juncaceae	*Juncus effusus*	Soft rush	[6-17]
Marantaceae	*Thalia dealbata*	Alligator-flag	[16-23]
Marantaceae	*Thalia geniculata*	Alligator-flag	[1]
Poaceae	*Phragmites communis*	Common reed	[8-17]
Pontederiaceae	*Eichhornia crassipes*	Water hyacinth	[10-11-13]
Typhaceae	*Typha angustifolia*	Carrow-leaved cattail	[4-14-16]
Typhaceae	*Typha domingensis*	Southern cattail	[4-11]
Typhaceae	*Typha latifolia*	Broad-leaved cattail	[4-6-14]

4 Conclusions

The use of constructed wetlands, employing different species or combination of species, has been a versatile alternative and useful in the treatment of various wastewaters. However, research is needed to further explore the potential of each species, with a view to identifying which species or group of species could be most appropriate for each type of waste. Another line of research that deserves further highlight is the selection of new species in different continents, because, due to diversity of flora in existence, it is likely that many species with great potentials for treating effluents are still unknown.

5 Acknowledgements

The authors thank the German Federal Ministry for Economic Cooperation and Development (BMZ), German Academic Exchange Service (DAAD) and the Exceed SWINDON Project for the financial support for participating at this expert workshop September 2016 in Recife.

6 References

[1]   Anning, A. K., Korsah, P. E., Addo-Fordjour, P., Phytoremediation of wastewater with *Limnocharis Flava*, *Thalia Geniculata* and *Typha Latifolia* in constructed wetlands. Int. J. Phytorem., 2013, 15, 452-464.

[2]   Appenroth, K. J., Krech, K., Keresztes, Á., Fischer, W., Koloczek, H., Effects of nickel on the chloroplasts of the duckweeds *Spirodela polyrhiza* and *Lemna minor* and their possible use in biomonitoring and phytoremediation. Chemosphere, 2010, 78, 216-223.

[3]   Bowmer, K. H., Nutrient removal from effluents by an artificial wetland: Influence of rhizosphere aeration and preferential flow studied using bromide and dye tracers. Water Res., 1987, 21, 591-599.

[4]   Brasil, M. da S., Matos, A. T. De, Soares, A. A., Plantio e desempenho fenológico da taboa (*Thypha* sp.) utilizada no tratamento de esgoto doméstico em sistema alagado construído. Eng. Sanit. Ambient., 2007, 12, 266-272.

[5]   Busnardo, M. J., Gersberg, R. M., Langis, R. Sinicrope, T. L., Zedler, J. B., Nitrogen and phosphorus removal by wetland mesocosms subjected to different hydroperiods. Ecol. Eng., 1992, 1, 287-307.

[6]   Coleman, J., Hench, K., Garbutt, K., Sexstone, A., Bissonnette, G., Skousen, J., Treatment of domestic wastewater by three plant species in constructed wetlands. Water, Air, Soil Pollut. 2001, 128, 283-295.

[7]   Corrêa, M. R., Velini, E. D., Arruda, D. P., Teores de metais na biomassa de *Egeria densa*, *Egeria najas* e *Ceratophyllum demersum*. Planta Daninha, 2002, 20, 45-49.

[8]   Gersberg, R. M., Elkins, B. V., Lyon, S. R., Goldman, C. R., Role of aquatic plants in wastewater treatment by artificial wetlands. Water Res., 1986, 20, 363-368.

[9]   Kivaisi, A. K., The potential for constructed wetlands for wastewater treatment and reuse in developing countries: a review. Ecol. Eng., 2001, 16, 545-560.

[10]   Klumpp, A., Bauer, K., Franz-Gerstein, C., Menezes, M., Variation of nutrient and metal concentrations in aquatic macrophytes along the Rio Cachoeira in Bahia (Brazil). Environ. Int, 2002, 28, 165–171.

[11]   Maine, M. A., Suñe, N., Hadad, H., Sánchez, G,; Bonetto, C., Nutrient and metal removal in a constructed wetland for wastewater treatment from a metallurgic industry. Ecol. Eng., 2006, 26, 341-347.

[12]   Maleva, M. G., Nekrasova, G. F., Malec, P., Prasad, M. N. V., Strzałka, K., Ecophysiological tolerance of *Elodea canadensis* to nickel exposure. Chemosphere, 2009, 77, 392-398.

[13]   Mishra, V. K., Tripathi, B. D., Accumulation of chromium and zinc from aqueous solutions using water hyacinth (*Eichhornia crassipes*). J. Hazard. Mater., 2009, 164, 1059-1063.

[14]   Neralla, S., Weaver, R. W., Varvel, T. W., Lesikar, B. J., Phytoremediation and on-site treatment of septic effluents in sub-surface flow constructed wetlands. Environ. Technol., 1999, 20, 1139-1146.

[15]   Pivari, M. O. D., Oliveira, V. B., Costa, F. M., Ferreira R. M., Salino, A., Macrófitas aquáticas do sistema lacustre do Vale do Rio Doce, Minas Gerais, Brasil. Rodriguésia, 2011, 62, 759-770.

[16]   Sohsalam, P., Sirianuntapiboon, S., Feasibility of using constructed wetland treatment for molasses wastewater treatment. Bioresource Technol., 2008, 99, 5610-5616.

[17]   Tanner, C. C., Plants for constructed wetland treatment systems — A comparison of the growth and nutrient uptake of eight emergent species. Ecol. Eng., 1996, 7, 59-83.

[18]   Vymazal, J., Constructed wetlands for wastewater treatment: five decades of experience. Environ. Sci. Technol., 2011, 45, 61-69.

[19]   WEF - Water Environmental Federation. Pretreatment of industrial waste: Manual of practice. WEF Alexandria/USA. 1994.

[20]   Wu, S.; Kuschk, P., Brix, H., Vymazal, J., Dong, R., Development of constructed wetlands in performance intensifications for wastewater treatment: a nitrogen and organic matter targeted review. Water Res., 2014, 57, 40-55.

[21]   Yang, Q., Chen, Z., Zhao, J., Gu, B., Contaminant removal of domestic wastewater by constructed wetlands: effects of plant species. J. Integr. Plant Biol., 2007, 49, 437-446.

[22]   Zhang, C, Wang, J., Liu, X., Zhu, S., Ge, H., Chang, S. X., Chang, J., Ge, Y., Effects of plant diversity on microbial biomass and community metabolic profiles in a full-scale constructed wetland. Ecol. Eng., 2010, 36, 62-68.

[23]   Zhao, F., Xi, S., Yang, X., Yang, W., Li, J., Gu, B., He, Z., Purifying eutrophic river waters with integrated floating island systems. Ecol. Eng., 2012, 40, 53-60.

ASSESSING ENVIRONMENTAL IMPACTS OF WASTEWATER TREATMENT ALTERNATIVES FOR SMALL SCALE COMMUNITIES

[1]Mustafa Yıldırım, [2]Bülent Topkaya

[1]Antalya Water and Wastewater Administration, General Directorate (ASAT), Antalya, Turkey, (yildirim.mustafa@gmail.com)

[2]Akdeniz University, Department of Environmental Engineering, Antalya, Turkey, (btopkaya@akdeniz.edu.tr)

Keywords: Constructed wetland, land treatment, LCA, stabilization pond, wastewater treatment.

Abstract

The effluents of wastewater treatment plants from small sized communities of less than 2000 population equivalents (p.e.), which are discharged into sensitive receiving water environments, must receive "appropriate treatment" according to the EU Urban Wastewater Treatment Directive. Appropriate treatment depends on the quality objectives of the receiving waters as well as the relevant provisions of the member states. In this study, wastewater treatment options, such as vegetated land treatment (VLT), constructed wetlands (CW), stabilization ponds (SP) and activated sludge treatment (AST), by which effluents are discharged to sensitive and less sensitive areas are evaluated by the life cycle assessment (LCA) approach. For this purpose, data related to energy usage, land requirement, raw material consumption, and released emissions from the life phases were collected with an inventory study and the environmental impacts were assessed by using SimaPro 7.1 LCA software. The results obtained from the assessments were compared with each other, which indicated that for small scale communities Vegetated Land Treatment and Constructed Wetlands are the most environmentally friendly wastewater treatment options.

1 Introduction

For the wastewater produced in communities of less than 2000 p.e., appropriate treatment is foreseen in case of discharges to sensitive water environments according to EU-Urban Wastewater Treatment Directive (UWWT) [1]. In recent years due to economic concerns, the municipalities tend to low-cost and energy saving systems. In this context eco-technological and economical solutions, such as natural systems, attain increasing interest. In addition to technical/financial parameters, environmental impacts of processes with similar performance should also be taken into consideration. Environmental impacts of different systems can be evaluated by the life cycle assessment (LCA) procedure, which is solely developed for this purpose [2, 3, 4, 5].

In this study, vegetated land treatment, constructed wetlands, rotating biological contactor, and conventional activated sludge treatment options, by which effluents are discharged to sensitive and less-sensitive areas, are evaluated by Life Cycle Assessment (LCA) approach. For this purpose, data related to energy usage, land requirement, raw material consumption, and released emissions from the life phases were collected with an inventory study and the environmental impacts were assessed by using SimaPro 7.1

2 Materials and Methods

2.1 Characteristics of wastewater

The characteristics of raw wastewater as well as the discharge criteria are assumed to be the same for all alternatives (Table 1).

Table 1: Characteristics of wastewater [1]

Parameter	Influent (mg/L)	Discharge criteria according to UWWT Directive (mg/L)	
		Less sensitive areas	Sensitive areas
COD	750	125	125
BOD_5	350	25	25
TSS	150	35	35
Tot P	10	-	1 - 2
Tot N	40	-	10 - 15

2.2 Wastewater treatment alternatives

2.2.1 Vegetated Land Treatment (VLT)

VLT system consists of a pre-treatment unit that includes a primary sedimentation unit for separating the solid material in wastewater and a vegetated area, where water can infiltrate, evaporate, and be accumulated by the plants [6, 7]. Pollutant removal in the VLT occurs during the flow of wastewater through the plant root/soil matrix (Fig. 1).

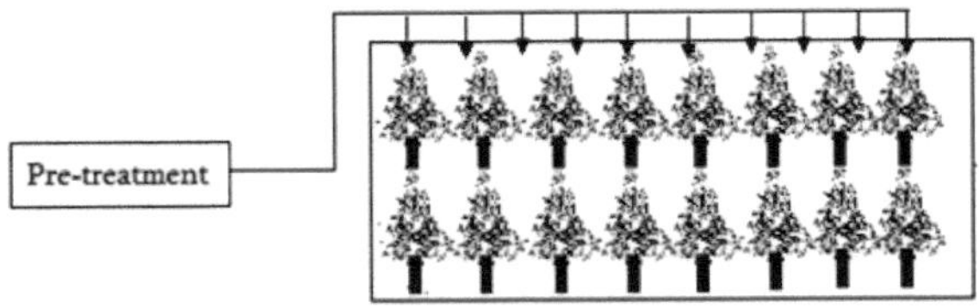

Figure 1: Vegetated Land Treatment

2.2.2 Constructed Wetland (CW)

CW is the other natural system evaluated in this study. Since 1950s, CWs have been used for reclamation of anthropogenic discharges such as wastewater, urban and rural runoff, sewage treatment, and mining activities because of their nutrient capturing capacity, simplicity, low construction/operation/maintenance costs, low energy demand, process stability, little excess sludge production, effectiveness, and potential for creating biodiversity [8, 9] (Fig. 2).

Figure 2: Constructed Wetland

2.2.3 Stabilization Ponds (SP)

Stabilization ponds are one of the simplest solutions for wastewater treatment which are particularly popular for small and medium-sized settlements [9, 10]. Stabilization ponds are a system that mimics nature and, therefore, does not need serious mechanical equipment and business skills (Fig. 3). However, the reaction rate is slow due to lack of any interference to the system. Therefore, there is a need for large land. Biological activity is susceptible to climatic conditions, in case climatic conditions vary a lot, and the effluent quality will be subject to fluctuations. In addition, odor and insect problems, and possible groundwater contamination are among the main disadvantages [9, 10]. Facultative ponds are the most commonly used type of stabilization ponds.

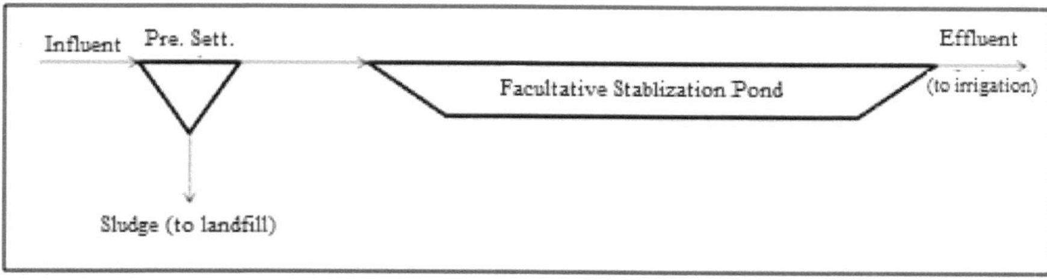

Figure 3: Stabilization pond

2.2.4 Activated Sludge Treatment Systems (AST)

AST is a well-known and commonly used biological treatment method for municipal wastewater. AST can be defined as the aeration of wastewater combined with organisms in order to develop microbiological flocs, in which the organic constituents in wastewater are utilized. Treatment of wastewater is completed after settling the flocs in the final clarifier and disinfection of effluent. In case of discharging the effluent to a sensitive area, it may be necessary to add a chemical treatment unit for the removal of nutrients. Additional information about activated sludge systems can be found in Metcalf and Eddy (2003) [9] (Fig. 4).

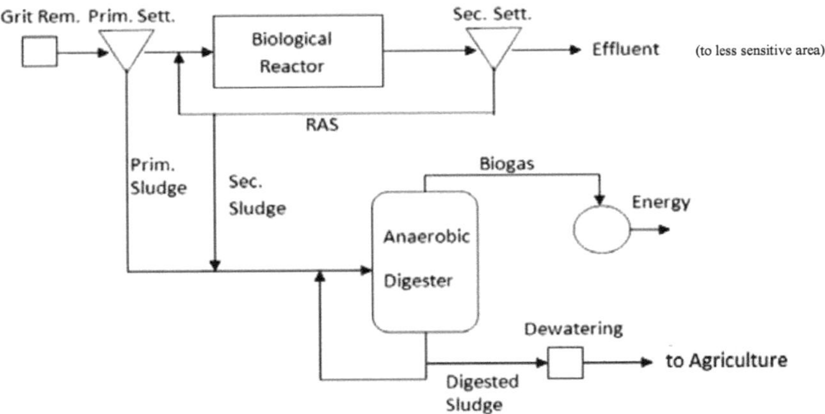

a)   Without phosphorus removal

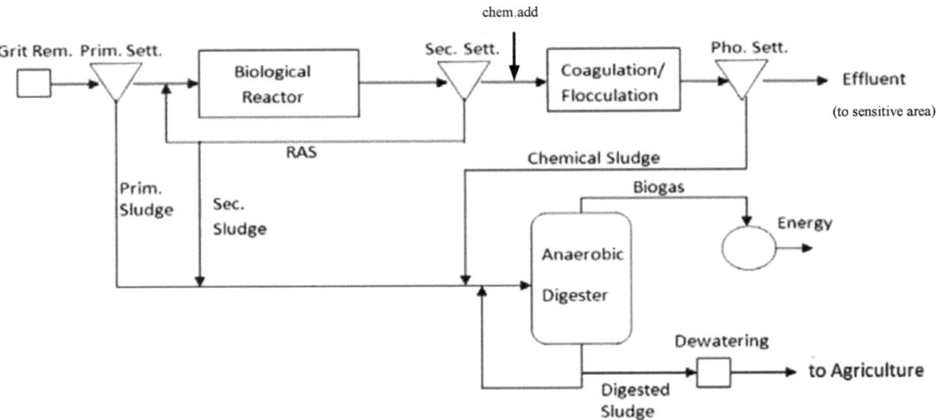

b)   With phosphorus removal

Figure 4: Flow diagrams of Activated Sludge System

2.3 LCA Procedure

In this study, the environmental performances of treatment systems are compared by using SimaPro 7.1 LCA software. The inventory data for treatment alternatives evaluated in this study are obtained from the EcoInvent V.2 database, in which data from various treatment systems exists. For the impact assessment phase CML baseline 2000 is used, which was developed by the Centre of Environmental Science at Leiden University. The data derived from the EcoInvent V.2 database are processed with CML baseline 2000 method and implicated by SimaPro 7.1 software.The details of the inventory data and assumptions regarding the treatment alternatives developed in this study are summarized below:

- VLT serves a population equivalent of 100 p.e. based on a specific water consumption of 210 L/p.e./day; the flow rate is calculated as 21 m^3/d. *Eucalyptus camaldulensis,* which is well adapted to wastewater conditions and has high nutrient uptake capacity, is chosen as biomass [4, 11].

- The free water surface type CW system also serves 100 p.e., and the flow rate is 21 m^3/d. Macrophytes in the wetland systems are selected as *Phragmites australis* [4]. It is assumed that these plants will be harvested every five years and disposed of on a sanitary landfill. Annual growth of the macrophytes is assumed as 3 kg/m^2 and annual disposed green waste is calculated as 28 kg/cap [12].

- Data related to stabilization ponds (SP) are obtained from in-situ measurements conducted by the facility located in Harran University Campus, which serves 190 p.e. The flow rate is 40 m^3/day and its volume is 864 m^3. COD concentration is daily measured in an eight months period in the facility, and heavy metal concentrations are measured in ASAT laboratory. Released CH_4 is calculated according to Ferrer (2002) and it is determined as 4.432 g CH_4/day.cap [13]. It is assumed that sludge is stabilized in the SP. According to Arceivala (2006), produced stabilized sludge in SP is 0.03 m^3/year.cap and the amount of stabilized sludge is calculated as 0.6 m^3/cap for the last 20 years' period [14].

- In the framework of this study, two alternative AST systems are investigated. The flow schemes of these systems are adapted from EcoInvent V.2 database. The first alternative is a conventional AST system consisting of mechanical and biological treatment units including sludge digestion. In this alternative, biogas obtained from the digesting unit is used for producing electricity energy. Stabilized sludge in the digester unit is applied to agricultural soils. Wastewater effluent is discharged to a less sensitive area.

  In the second alternative, nutrient removal is aimed prior to discharging of the treated water to sensitive areas. In order to comply with the requirements of the UWWT Directive, a chemical phosphorus precipitation unit is added.

In the framework of the impact assessment study, the emission data obtained from the inventory step are sorted into categories according to their effects. In order to evaluate the effects, these sorted emissions are converted to same dimension using characterization factors. In this study, the characterization factors and impact categories are selected according to CML2 baseline 2000, which is a life cycle assessment method included in SimaPro7.1 software. This method is one of the few which includes characterization factors for the emissions resulting from organic material, phosphorus and nitrogen containing compounds

[15]. The default impact categories included in this method are shown in Table 2. These categories can be selected according to the characteristics of each study. In the framework of this study, all default categories are taken into consideration.

3 Results and Discussion

In order to differentiate the environmental impacts of the wastewater treatment options, life cycle impact assessment using the inventory data (LCI) mentioned above is conducted. In this context, the contributions of each treatment alternative to impact categories are calculated in the normalization step. Normalization provides a basis for comparing different types of environmental impact categories, in which all impacts get the same unit. In this manner, it was possible to compare all environmental impacts of the treatment options. In the normalization step, local specifications are taken into account for the evaluation of the local-dependent impact categories (eutrophication and toxicity categories). World normalization data listed in CML2 baseline 2000 are used as specific data for these impacts [15]. Normalization results indicate the categories, which are most influenced by the treatment options (Table 7). In all categories except of eutrophication, the AST system with P-removal option has the highest impacts. Global warming potential (GWP) for the VLT option is negative in accordance with the literature [3]. Release of emissions and activities (such as raw material usage) affecting the other factors mainly occurred in the construction phase, and environmental impacts resulting from construction are relatively low taking into account the time period.

Table 2: Released emissions on the impact categories of the treatment alternatives per capita in a 20 years' period.

Category*	Unit**	VLT	CW-LS	CW-S
AD	kg Sb eq	0.05	1.07	1.69
A	kg SO_2 eq	0.50	0.72	1.13
E	kg PO_4^{3-} eq	9.74	18.21	12.45
GWP	kg CO_2 eq	-1,850	633	1,008
T	kg 1,4-DB eq	565	1,1491	18,209

Category*	Unit**	SP	AST-LS	AST-S
AD	kg Sb eq	1.95	0.44	1.87
A	kg SO_2 eq	1.44	3.79	4.67
E	kg PO_4^{3-} eq	16.16	31.70	19.20
GWP	kg CO_2 eq	1,472	405	672
T	kg 1,4-DB eq	20,731	44,107	145,306

*AD: Abiotic Depletion, A: Acidification, E: Eutrophication, GWP: Global Warming Potential, T: Toxicity
** Sb eq: Antimony equal, SO_2 eq: Sulfurdioxide equal , PO_4^{3-} eq: Ortophosphate equal, CO_2 eq: Carbondioxide equal, 1,4-DB eq: 1,4-Dichlorobenzene equal

4 Conclusions

LCA is a powerful tool and it helps decision makers to evaluate the different treatment systems according to their environmental performance. There are several studies dealing with the environmental impacts of wastewater treatment systems. In this study, treatment plants suitable for small-scaled residential centers are evaluated. For small scale communities with <2000 p.e., there are several treatment options. These systems will gain importance in the near future. It can be concluded that the decision on the best available treatment option requires detailed knowledge about the requirements of the receiving environment, the environmental impacts and the economic conditions of the option. It is essential that the systems used in this study treat wastewater with same properties, and the effluent quality fulfils the same standards. Thus, it is easier to compare the systems in a more detailed manner.

Based on the calculated environmental impacts, showed in Table 7, the vegetated land treatment system (VLT) is the most appropriate option. It results in the lowest environmental impacts for all categories. The GWPs of the VLT systems are negative as the biomass used in the system assimilates CO_2 while plant growth. However, VLT option is not applicable in large capacities since area requirement per capita is very high (about 20 m^2). In addition, this system is not applicable in closed basins where groundwater is vulnerable to contamination.

The constructed wetland systems (CW) are more feasible than VLT in regions with higher groundwater vulnerability as they have an impermeable bottom seal, which is composed of consolidated clay and an HDPE layer. Additionally, a receiving (water) environment is needed to discharge the treated wastewater. CW has a higher GWP than VLT. The main drawback is also the relatively high requirement of land. By discharges to less sensitive areas, the specific land requirement is calculated as 4-6 m^2/p.e./year. The required area increases to 8-10 m^2/p.e., in case the effluent is discharged to a sensitive receiving area.

The other option, the activated sludge treatment systems (AST) with and without phosphorus removal, has the highest environmental impacts in construction and operation phases, compared with the natural systems. On the other hand, AST systems have the lowest area requirements, and the specific area requirements decrease with increasing capacity.

The last alternative, the stabilization ponds (SP) have long hydraulic retention time (HRT). The phosphorus removal is good due to both phosphorus precipitation resulting from long HRT and algal uptake. SP behaves facultative, and anaerobic conditions occur in the bottom. Therefore, CH_4 emission is high although a part of CH_4 is consumed throughout the water column. But, most of the organic matter is mineralized and CO_2 is emitted.

Generally, land requirements of the treatment alternatives are a major factor, which is the case especially for coastal residential centres. This issue can play a key role on the final decision making. For small scale communities' with less than 2000 p.e. and effluent discharges to less-sensitive environments, VLT and CW are the most environmentally friendly options. For these communities, AST can replace the natural treatments in case that not enough land is available. By discharges to sensitive areas from these communities with <500 p.e., VLT and CW are likely the best option. For greater capacities, AST should be preferred.

5 Acknowledgements

This manuscript is produced from a research project supported by The Scientific Research Projects Coordination Unit of Akdeniz University (Project number: 2010.03.0121.006).

The authors want also to thank the Exceed Swindon Project, which enabled the participation at this workshop.

6 References

[1]   EU Council Directive 91/271/EEC. ''Urban Wastewater Treatment Directive.   http://eur-lex.europa.eu/legal-content/EN/TXT/PDF/?uri=CELEX:31991L0271&from=EN

[2]   A. K. Søvik, J. Augustin, K. Heikkinen, J. T. Huttunen, J. M. Necki, S. M. Karjalainen, B. Kløve, A. Liikanen, U. Mander, M. Puustinen, S. Teiter, P. Wachniew. Emission of the greenhouse gases nitrous oxide and methane from constructed wetlands in Europe, *J. Environ. Qual.* 2006, 35, 2360–2373.

[3]   A. A. Jensen, L. Hoffman, B. T. Møller, A. Schmidt, K. Christiansen, J. Elkington, F. Van Dijk, *Life Cycle Assessment (LCA): A guide to approaches, experiences and information sources*, Report to the Environmental Agency, Environmental Issues Series, No. 6. Copenhagen 1997.

[4]   A. P. Machado, L. Urbano, A. G. Brito, P. Janknecht, J. J. Salas, R. Nogueira, Life Cycle Assessment of wastewater treatment options for small and decentralized communities. *Water Sci. Technol.*, 2007, 56 (3), 15–22.

[5]   Pre´ Consultants. *"SimaPro: Introduction to LCA with SimaPro"*. PRe´ Consultants, Amersfoort, The Netherlands 2006.

[6]   M. Madison, M. Henderson, *Zero discharge all-weather land application with soil storage*. Proceeding of the 66th Annual Conference of the Water Environment Federation, Anaheim, CA, 1993.

[7]   A. P. Kruzic, Natural treatment systems, *Water Environ. Res.* 1994, 66/4, 357-361.

[8]   E. A. Kaygusuz. *Domestic wastewater treatment in pilot-scale constructed wetlands implemented in the Middle East Technical University*, PhD Thesis, Natural and Applied Science Ins., METU, Ankara, Turkey, 2004.

[9]   Metcalf & Eddy, Inc. *Wastewater engineering, treatment and reuse*, McGraw Hill Publishing, 4th Edition, 2003.

[10] S. R. Qasim, Wastewater treatment plants: Planning, design, and operation, CRC Press, 1999

[11] M. Bhati, G. Singh, Growth and mineral accumulation in *Eucalyptus camaldulensis* seeding irrigated with mixed Industrial effluents. *Biosource Technol.* 2003, 88, 221-228.

[12] J. Vymazal, L. Kröpfelova, Growth of *Phragmites australis* and *Phalaris arundinacea* in constructed wetlands for wastewater treatment in the Czech Republic, *Eco. Eng.*, 2005, 25, 606-621.

[13] A. V. M. Ferrer, Algae and duckweed-based pond reactors: An evaluation and comparison of CH_4 gas emissions. 2002, MSc. Thesis Institute Unesco-Ihe, Delft, p. 42

[14] S. J. Arceivala, Wastewater treatment for pollution control and reuse. Mc Graw Hill Publishing, 3rd Edition. 2006, ISBN-13: 978-0070620995.

[15] J. B. Guinee (Ed), *Life Cycle Assessment, an operational guide to the ISO standards*, Final Report, Centre of Environmental Science, Leiden University (CML), 2001.

EVALUATION OF DECENTRALIZED APPROACHES TO WASTEWATER TREATMENT SYSTEMS AS ALTERNATIVE TO ENVIRONMENTAL IMPACTS AND INDICATION OF SUSTAINABLE SOLUTION FOR WATER CONSERVATION IN CITIES

Mariana Cardoso Chrispim

University of Sao Paulo, Graduate Program in Sustainability, Av. Arlindo Bettio 1000, Sao Paulo, SP, 03828-000, Brazil (mariana_chrispim@hotmail.com)

Keywords: Nutrients, Reuse, Sanitation, Sustainability, Water

Abstract

The Millennium Development Goals aim to achieve the sustainable development and to contribute improving access to water supply, sanitation, and energy, and to reduce hunger in the world. Energy and water are essential resources for food production, however, are becoming scarce. In this context an integration of the complete water management cycle includes water conservation and reclamation, storage of reclaimed water and rainwater, and wastewater treatment mainly in decentralized models. Wastewater treatment systems can generate products such as: renewable energy, fertilizers and water. This study focuses on the relations between wastewater treatment systems and its environmental impacts, besides to indicate potential solutions. In addition, it is also presented results from an experimental greywater treatment system operated at the university. The objective of this study was to address some Brazilian experiences of centralized wastewater treatment plant (large scale) and decentralized wastewater treatment plant (small scale). We expect that the results of this study can offer a review of strategies of wastewater treatment to implement in urban areas in Brazil, and to create subsidies for the planning of new wastewater treatment plants with low environmental impacts, including sustainable solutions (e.g., water reuse).

1 Introduction

Human beings use the natural resources mainly for purposes of food production, energy and water supply. Land use, urbanization increase, population growth and the technologies in these processes are factors that affect the quality of these natural resources. In this paper, the focus is set on water conservation strategies experienced in Brazil, including greywater treatment and reuse, rainwater harvesting, collection of water from air conditioner, and finally some wastewater treatment plants, which have sustainable practices and technologies, are going to be briefly described.

Greywater, the household wastewater produced by showers, baths, lavatories, kitchen sinks and laundries, comprises 50-80% of total domestic wastewater [1]. Greywater reuse is an important practice because it can reduce municipal wastewater production and lower demand for potable water. A Brazilian study (from 2007) [2] estimated that the greywater reuse may generate an economy of 25-30% of potable water consumption in a household. Currently, potable quality water is used for the most part, not just for drinking (which is only about 2-4 L per capita per day), but also for washing, flushing toilets, and irrigation in agriculture (among others). The reuse of treated wastewater for non-potable purposes could significantly reduce the overall societal water footprint [3]. Consequently, greywater reuse has the potential to decrease urban wastewater discharge and to improve environmental health by reducing environmental impacts such as energy consumption, and water and land pollution. Chemical, physical and biological treatment processes have been evaluated for treating greywater [4].

In this study, a biological process called moving bed biofilm reactor in a university campus was evaluated and the performance and removal efficiencies monitored. The MBBR is a biological process based on biofilm attached growth in order to improve efficiencies for organic matter degradation and nutrient removal. The technology was developed in the late 1980s by the Norwegian company Kaldnes Miljøteknologi and has been applied in municipal and industrial wastewater treatment plants in several countries [5]. Inside the reactor, there are free-floating carriers that provide protected surface area to support the growth of heterotrophic and auto-trophic bacteria. The main goal of the study was to evaluate an MBBR reactor at pilot scale for greywater treatment. This technology has some advantages such as: being compact, easy to maintain, and feasible to use in urban water systems with high population density and lack of available spaces, a common situation in large cities in developing countries. Some results are going to be presented in the results section. The performance of the evaluated treatment was sufficient to meet Brazilian water quality requirements for several types of irrigation, including agricultural, and it could be installed at decentralized scale to treat greywater of several households. Although this study considered one only building, this system could be installed at decentralized scale to treat greywater from more households.

The greywater reuse as an alternative water source is particularly interesting for Metropolitan Region of Sao Paulo, region with about 20 million inhabitants and one of the largest industrial complexes in the world. Sao Paulo has an insufficient flow to supply the water demand of the city's largest metropolitan region in the country, located in the state of São Paulo. The state of São Paulo, where 22% of the country's population lives, has just 1.6% of the surface water available in the country [6]. In 2014, Sao Paulo registered the worst drought in eight decades in southeastern Brazil. Water levels dropped dramatically in the reservoirs that supply São Paulo, the country's largest city. Inhabitants of São Paulo reported that their water supplies have been restricted. The Government created measures to try solving the problem and promoted discount for users, who reduced their water consumption. This factor and the population growth increase the water demand [7].

Another sustainable water management strategy is rainwater harvesting. This can contribute as an alternative water source, consequently promoting potable water savings of about 33% [2]. In addition, during heavy rains, the rainwater harvesting decreases the surface water that drains to the public sewer. Figure 1 shows an example of rainwater tank installed at the State University of Mato Grosso do Sul, Brazil.

Figure 1: Rainwater tank at the university (A) and filter before rainwater tank (B).

Rainwater collection system was built at the State University of Mato Grosso do Sul, located in the city of Coxim-MS, with a capacity to store 400 L of water. The two tanks were connected to each other through a hose (communicating vessels). Once a week, 40 mL of chlorine was added to the rainwater tank according to IPT [8]. Some safety recommendations for rainwater harvesting include appropriate maintenance and filtration because water can include some contaminants such as algae, air pollution, bird excrements, leaves, sand, and dust. The installation of filters can remove these contaminants and prevent mosquitoes breeding. A filter (wire mesh) was installed according to SEMPRESUSTENTÁVEL instructions [9]. After storage this water can be used for toilet flushing, cleaning, vehicle washing, and irrigation.

In addition to rainwater harvesting, another source in households is air conditioner water. Frequently, the condensation produced by most air conditioner systems is drained into the sewer, and the water is lost. This water is formed from condensed water vapor. The collected water can be used for cleaning and irrigation. The city of Coxim has an average temperature of 26 °C, and most of the buildings (residential and commercial) have air conditioners. So we started to collect water from 10 air conditioners (total of about 70 L per day) at the State University of Mato Grosso do Sul and used it in irrigation of green areas of the campus. A study developed by another Brazilian university evaluated the quality of air conditioning water and concluded that the physicochemical parameters (hardness, pH, alkalinity, conductivity and chloride values) met the requirements for potable water of the Ministry of Health [10].

After discussing the water demand, it is necessary to address how water is going to be treated after use. Wastewater treatment plants cause environmental impacts such as greenhouse gas (GHG) emissions (carbon dioxide CO_2, methane CH_4, and nitrous oxide N_2O), material and energy consumption, and generation of sludge. Currently, there are some studies that compare different technologies of wastewater treatment and their environmental impacts [11]. Source-separation wastewater technologies include, for example, separation of greywater, urine, and feces. The separate collection of human urine reduces nutrients loading to domestic wastewater and facilitates nutrient recycling as well [12]. Therefore, currently there are possible alternatives for reducing the environmental impacts in construction and operation (of wastewater treatment plants like energy consumption for treatment steps, manufacture of chemicals and maintenance material, resources consumption, and handling of process-related emissions and waste, etc.. These alternatives include energy recovery (ex. Biogas) and use the treated water for other activities [13].

In Brazil, most wastewater treatment plants dispose of their excess sludge in landfills, which can cause air, water and land pollution [14]. In addition, only 48.6 % of population in Brazil has access to sewage collection and 40% of sewage of the country is treated [15]. This could be a result of centralized approach, which includes the installation of large wastewater treatment plants. The impacts of inadequate sanitation are water and soil pollution, and the increase the prevalence of water-related diseases.

The decentralized wastewater treatments are defined as systems where the distance from the wastewater source, its treatment and disposal are close to each other, not needing long sewage collection pipes [16]. Decentralized wastewater treatments systems can be a solution mainly to serve population from small and medium cities, which do not have a sewage collection, as well as to isolated communities or low income households, increasing the access to sanitation with low costs [17].

In addition to the greywater treatment results, this paper also presents a review of various existing approaches to wastewater treatment and management, and discusses decentralized sanitation systems as possible solutions to reduce environmental impacts.

2 Materials and Methods

2.1. Greywater pilot plant set-up

The greywater sources in the staff building were four showers, two washbasins and one washing machine situated in a building with two changing rooms used by staff at the University of Sao Paulo, Brazil. The pilot scale greywater treatment unit was built outside of the staff building and consisted of three plastic collection tanks, an equalization basin, an aerobic moving bed biofilm reactor (MBBR), and a settling tank. The raw greywater from these sources were characterized. Greywater was collected and transported by gravity to the subsurface equalization basin, where it was pumped to the MBBR using a ProMinent/Vario C pump (maximum flow rate: 31.2 L/h). The MBBR contained carrier elements and aerators (air compressor Boyu model ACQ-008; maximum air flow: 110 L/min). The operational conditions of the reactor are described on Table 1 [18].

Table 1: Operational conditions of the MBBR reactor

Parameter	MBBR reactor	Unit
Daily flow rate	302	L/d
Hydraulic Retention Time	4	h
Volume of reactor	83.3	L
Filling ratio	14	%
Carrier surface area	490	m^2/m^3

Synthetic greywater was used to evaluate the performance of the experimental greywater treatment system. It was prepared according to NSF/ANSI 350 [19], but without the addition of secondary effluent. The MBBR was in continuous operation for three months for treatment of synthetic greywater. During this period, samples were collected from the following points: equalization basin effluent (raw), inside the MBBR reactor, and after the settling tank (treated). Samples were analyzed for turbidity, TSS, pH, DO, EC, oils and grease, anionic surfactants, total N, total P, COD, BOD, TOC. Potential applications for treated water were identified based on final water quality, in consultation with national and international regulations for non-potable water reuse [18].

3 Results and Discussion

3.1. Greywater characterization and treatment results

Greywater quality can vary due to many household factors including products used in the household, building infrastructure, and residents' behavior, activities, income and age. Generally, greywater contains low total suspended solids, high turbidity, and high concentrations of phosphorus, surfactants, oils and greases. Greywater can also be polluted with microorganisms due to excreta contamination, residues from skin, and vomit, but the contamination with fecal coliforms is generally low [20]. It is possible to note that greywater from showers has the highest *E. coli* concentration (5.06×10^4) [18]. This was stated by previous studies [21, 22]. The results of microbial analyses of raw greywater showed: 7.6×10^6 CFU per 100 mL for lavatories; 4×10^5 for showers and 50 for washing machine [18]. For most physicochemical parameters (Table 2) the results of greywater characterization were in the range reported in the literature. Oil and grease and surfactants concentrations are important parameters to monitor, if the treated water will be used in irrigation because it may affect negatively the soil properties and consequently the plant growth. In our study, greywater from kitchen was not considered, therefore, the oil and grease concentration was low in raw greywater of all sources (22.5 mg/L); the highest concentration was in washing machine samples then in bathroom samples.

According to Brazilian standard 357 [23], the maximum allowable concentration of anionic surfactant for discharging into freshwater is 0.5 mg/L. This means that the greywater must be treated before being discharged in surface water. In addition to concerns related to reuse, the excess of surfactants in biological treatments may cause operational problems such as inhibition of microorganism growth and adhesion to the reactor surface. We could observe hairs and clothes lint from the raw greywater on the pretreatment of fine screen (1 mm).

Table 2: Characteristics of raw greywater from each source [18]

Parameter	Showers	Lavatories	Washing machine
Turbidity (NTU)	100	48.7	33.5
TSS (mg/L)	156	18.5	32.7
TS (mg/L)	408	165	901
COD (mg/L)	273	208	274
BOD (mg/L)	123.1	101	77
Total-P (mg/L)	5.3	3.3	2.3
Total-N (mg/L)	50.3	5.1	4.3

The results indicated that TSS, COD and turbidity in this study were lower than those measured for kitchen greywater [24]. Biofilm development on the carriers was observed, especially on their internal surfaces. However, biofilm layer was thin as compared to previous studies [25]. The limited nutrient content and low organic matter concentration in raw greywater from our experiment might have resulted in slower biomass growth. During monitoring the pilot treatment for synthetic greywater, some results of removal efficiencies were: 66% of turbidity; 87% of TSS; 12% of Total P; 59% of BOD; 70% of COD; 30% of Anionic Surfactants; and 25.8% of oils and grease [18]. In comparison with other studies that evaluated biological processes for greywater treatment (e.g. membrane bioreactor, rotating biological contactor [25]) we measured lower removal efficiencies of Total P, BOD and turbidity. But it is important to note that the operational conditions vary between the studies (e.g., HRT, influent characteristics) which affect treatment performance. The final effluent did not achieve the requirements for indoor uses because of high turbidity. According to Brazilian standard NBR 13969 [16], the turbidity of final effluent did not comply with the limits set for toilet flushing, floor cleaning, or ornamental purposes. However, the final effluent did meet the requirements for reuse in irrigating orchards, pastures, cereals and other crops through surface flood or drip irrigation [16].

Preliminary tests of the pilot-scale system have been conducted with real versus synthetic greywater. Preliminary results demonstrated that, while the removal efficiencies of some parameters were higher (75% turbidity, 78% COD, 75% TOC and 97% total coliforms) for real greywater than synthetic greywater treatment, the final effluent quality was similar for both [18]. In addition to the treatment performance discussed previously, other aspects such as the investment costs, maintenance needed as well as the suitability of the treatment system for on-site applications should be studied for a better decision making. Compared with waste-water reuse, greywater recovery involves smaller hygienic concerns and so less treatment efforts [12].

3.2. Sustainability in decentralized and centralized wastewater treatment plants
Decentralized systems keep the collection component of the wastewater management system as minimal as possible and focus mainly on necessary treatment and disposal of wastewater. The objectives of wastewater treatment plants are to protect public health and the environment, and to reduce pressure on scarce water resources. In this context, decentralized system has the characteristic of no use of water as a transportation medium [26]. For rural communities, the conventional centralized system is not only expensive in terms of provision of services (e.g., pumping costs), but operation and maintenance as well. The advantages of decentralized systems are that total costs for operation and maintenance are lower than of centralized system, so it is a good option for rural areas [26].

One example of decentralized household sewage treatment in Sao Paulo city was an experimental research: a pilot sewage treatment station, built on the Hydraulic Technological Center of University of Sao Paulo (USP), in São Paulo. The raw sewage was coming from the residential condominium of the USP and the central restaurant of the University City. The average and maximum flows of raw sewage of 640 L/d and 1600 L/d, respectively, have been applied to the septic tank input to 5,000 L and drained by gravity from the edge of the hybrid constructed wetland with TDH total of 2.8 d and 1.1 d, respectively, under application of the average and maximum flow rates. Monitoring the experiment in the field lasted 6 months. The physico-chemical and microbiological parameters of raw sewage, effluent from septic tank, and the chambers of hybrid constructed wetland were evaluated during 97 consecutive days. The results indicated that both the young seedlings as well as the adults of Vetiver grass have adapted well to environmental conditions. The average removal efficiencies in the final treated effluent as to carbonaceous organic matter were 96% for BOD and 90% for COD, 40% for Total N, 60 % for Total P, 74% for SST, and 97 % for oils and greases. The average removal of thermo-tolerant coliform was of 2 and 3 log units, and average of *E. coli* 1 and 3 log units. The quality of the final treated effluent obtained in this study met the requirements and emission standards of liquid effluents in water bodies and in public sanitation systems defined in federal environmental legislation of the State of São Paulo in Brazil [27].

Wastewater contains important resources, like water, organic matter, heat, and nutrients such as phosphorus and nitrogen, which can be used. Wastewater Treatment Plants use great quantities of water and energy, generate effluents, emissions to atmosphere, and solid waste that need to be treated and disposed of [28]. The degree of these impacts depends on the influent characteristics, which is being treated, and the process used (physical, chemical or biological). Sustainable solution in wastewater treatment in cities includes nutrient and energy recovery, and water reuse. There are some examples in Brazil which are considering these technologies. In Brazilian state of Paraná, there is a project ETE Belem of cleaner energy production from digestion of wastewater sludge (biogas), generating 5.8 MW electricity. Of the energy gained, 0.5 MW is going to be used in the power plant and the rest is distributed via public network (equivalent to 2,100 households). This is a sustainable solution because otherwise the sludge would be disposed of in landfill leachate, consequently generating air pollution (if burned) or landfill leachate polluting land if not previously treated. This wastewater treatment plant already uses part of the treated effluent for non-potable purposes. The other product – bio-fertilizer - is applied in agriculture [29].

Another example is Aquapolo Project, Water Reuse System for Industrial Purposes. The effluent from ABC wastewater treatment plant is the source to supply industrial purposes (Petrochemical Complex). This wastewater treatment plant treats sewage from several cities (Santo André, São Bernardo, Diadema, São Caetano, Mauá and part of São Paulo). São Paulo state contains 645 municipalities. Companhia de Saneamento Basico do Estado de Sao Paulo (SABESP) is the water and sanitation company responsible for the services in this state. SABESP operates in 365 of these municipalities, in which it supplies 28 million people directly with treated water, 22.8 million with wastewater collection, and 14.7 million with wastewater treatment. In order to expand the services to all the State cities, new connections of water and sewerage are needed [14]. There are some industries in Brazil, which practice water reuse and energy recovery from anaerobic wastewater treatment facilities. This can reduce environmental risks (renewable fuel), for example, at Laticínios Bela Vista - production of cheese, butter, long life milk, condensed milk, and chocolate-flavored, low-fat dairy drink [30].

4 Conclusions

It is important to stimulate wastewater treatment plants with these solutions in Brazil because there are only few scientific studies involving nutrient removal or energy recovery from wastewater. Some international papers evaluated the nitrogen recovery from domestic wastewater through aquatic plants for biomass production. For source separation systems, struvite precipitation is widely used for nutrient recovery. The combination of greywater and rainwater are viable options to meet the water demand mainly for non potable purposes. Reuse of wastewater could indeed contribute to overcoming water shortage. Further research should be conducted to evaluate the feasibility implementing the nutrient recovery in a wastewater treatment plant in São Paulo city. This approach tries to extract more from the existing centralized plants, and to identify whether more benefits could be gained under real conditions.

5 Acknowledgements

We highly acknowledge the financial support of DAAD and Exceed Swindon Project to participate at this event, and the financial support from Funding Authority for Studies and Projects (FINEP) - MCT/MCIDADES/FINEP/Saneamento Ambiental e Habitação - 6/2010, and the Coordination for the Improvement of Higher Education Personnel (CAPES).

6 References

[1]   Nolde, E., Greywater treatment systems in Germany: Results, experiences and guideline. Wat. Sci. Tech., 2005, 5(10), 203-10.

[2]   Ferreira, D.F., Ghisi, E., Potential for potable water savings by using rainwater and greywater in a multi-storey residential building in southern Brazil. Buildin. Environ., 2007, 42(7),10.

[3]   Chandran, K., Transforming wastewater AMA, 2016, Available at: https://www.reddit.com/r/science/comments/4w4e4s/science_ama_series_im_kartik_chandran_and_im/

[4]   Li, Z., Boyle, F., Reynolds, A., Rainwater Harvesting and Greywater Treatment Systems for Domestic Application in Ireland. Desalin., 2010, 260(1-3), 1-8,.

[5]   Reis, G.G., Influência da carga orgânica no desempenho de reatores de leito móvel com biofilme (MBBR), (Organic load effect on the performance of a moving bed biofilm reactor (MBBR). MS thesis. Universidade Federal do Rio de Janeiro, Rio de Janeiro (in Portuguese), 2007.

[6]   Martirani, L.A., Peres, I.K., Crise hídrica em São Paulo: cobertura jornalística, percepção pública e o direito à informação. Ambient. Soc., São Paulo, 2016, 19(1), 1-20.

[7]   Chrispim, M.C., Avaliação de um sistema de tratamento de águas cinzas em edificação de campus universitário(Evaluation of a greywater treatment system in a building of a university campus). MS thesis, Universidade de São Paulo, São Paulo, 2014. doi:10.11606/D.6.2014.tde-11092014-102851.

[8]   Instituto de Pesquisas Tecnológicas (IPT), Manual para captação emergencial e uso doméstico de água de chuva (Handbook to emergency rainwater harvesting and domestic use), 2015.

[9]   Sempresustentável, Projeto experimental de aproveitamento de água da chuva com a tecnologia da minicisterna para residência urbana- Manual de construção e instalação, 2014, Available at: http://www.sempresustentavel.com.br/hidrica/minicisterna/minicisterna.htm

[10] Carvalho, M.T.C., Cunha, S.O., Faria, R.A.P.G., Caracterização quali-quantitativa da água da condensadora de aparelhos de ar condicionado (Quali-quantitative characterization of condesation water from air conditioner). III Congresso Brasileiro de Gestão Ambiental Goiânia, 2012.

[11] Matos, C., Pereira, S., Amorim, E.V., Bentes, I., Briga-Sa, A, Wastewater and greywater reuse on irrigation in centralized and decentralized systems - An integrated approach on water quality, energy consumption and CO2 emissions. Sci. Total Environ., 2014, 493, 463–471.

[12] Larsen, T.A., Hoffmann, S., Lüthis, C., Truffer, B., Maurer, M., Emerging solutions to the water challenges of an urbanizing world. Science, 2016, 352(6288), 928-933.

[13] Zang, Y., Li, Y., Wang, C., Zhang, w., Xiong, W., Towards more accurate life cycle assessment of biological wastewater treatment plants: a review. J. Cleaner Prod., 2015, 107, 676–692.

[14] Waterworld 2011. Brazil's water and sanitation trailblazer. Available at: http://www.waterworld.com/articles/2011/09/brazil-s-water-sanitation.html

[15] Sistema Nacional de Informações sobre Saneamento (SNIS). D agnóstico dos Serviços de Água e Esgoto-2012. Brasília: Ministério das Cidades/SNSA, 2014, 164.

[16] ABNT – Associação Brasileira de Normas Técnicas. NBR 13.969: Tanques sépticos – Unidades de tratamento complementar e disposição final dos efluentes líquidos – Projeto, construção e operação. Rio de Janeiro; 1997.

[17] Fiúza Jr, A.P.F., Philippi, L.S., Uma análise da gestão do saneamento descentralizado em município de médio porte–Estudo de caso: Blumenau-SC. 23º Congresso Brasileiro de Engenharia Sanitária e Ambiental, 2004.

[18] Chrispim, M.C., Nolasco, M.A., Greywater treatment using a moving bed biofilm reactor at a university campus in Brazil. J. Cleaner Prod., 2016, 142, Part. 1, 290-296.

[19] NSF/ANSI 350. Onsite residential and commercial water reuse treatment systems. NSF Standard for Wastewater Treatment Systems/American National Standard, Michigan, USA, 2011, 23.

[20] Jefferson, B., Palmer, A., Jeffrey, P., Stuetz, R., Judd, S., Grey water characterization and its impact on the selection and operation of technologies for urban reuse. Water Sci. Technol., 2004. 50(2), 157-64.

[21] O'Toole, J., Sinclaire, M., Malawaraarachchi, M., Hamilton, A., Barker, S.F., Leder, K., Microbial quality assessment of household greywater. Water Res., 2012, 46(13), 4301-4313

[22] Vakil, K.A., Sharma, M.K., Bhatia, A., Kazmi, A.A., Sarkar, S., Characterization of greywater in an Indian middle-class household and investigation of physicochemical treatment using electrocoagulation. Sep. Purif. Technol., 2014, (130), 160–66.

[23] Brazil, 2005. Conselho Nacional do Meio Ambiente (The National Environmental Council). Resolução nº 357 (Resolution 357). Establishes provisions for the classification of water bodies as well as environmental directives for their framework, establishes conditions and standards for effluent releases and makes other provisions (in Portuguese). 2005.

[24] Travis, M.J., Weisbrod, N., Gross, A., Accumulation of oil and grease in soils irrigated with greywater and their potential role in soil water repellency. Sci. Total Environ., 2008,.394, 68-74.

[25] Jabornig, S., Favero, E., Single household greywater treatment with a biofilm membrane reactor (MBBMR). J. Membr. Sci., 2013, 446, 277-285.

[26] Massoud, M.A., Tarhini, A., Nasr, J.A., Decentralized approaches to wastewater treatment and management: Applicability in developing countries. J. Environ. Manage. 2009, 90, 652–659.

[27] MENDONÇA, A.A.J. Avaliação de um sistema descentralizado de tratamento de esgotos domésticos em escala real composto por tanque séptico e wetland construída híbrida. MS thesis,Faculdade de Saúde Pública, Universidade de São Paulo, São Paulo, 2015.

[28] Lins, G.A. 2010. Impactos ambientais em Estações de Tratamento de Esgotos (ETEs). MS thesis, Universidade Federal do Rio de Janeiro. Rio de Janeiro, 2010.

[29] GE Imprensa Brasil, Estação de tratamento de esgoto gerará energia no Paraná. Available at: http://www.geimprensabrasil.com/estacao-de-tratamento-de-esgoto-gerara-energia-no-parana, Access on: 27 Aug. 2016.

[30] Santana, V., Estação de tratamento de rejeitos é a 1ª do país a gerar energia em laticínio. 2015. Available at: http://g1.globo.com/goias/noticia/2015/11/estacao-de-tratamento-de-rejeitos-e-1-do-pais-gerar-energia-em-laticinio.html, Access on: 25 Aug. 2016.

COMPARATIVE RELIABILITY ANALYSIS FOR EFFLUENT QUALITY OF CENTRALIZED AND DECENTRALIZED WASTEWATER TREATMENT PLANTS

[1,2]**Juliana C. Morais**, [3]**Elizabeth Pastich**, [2]**Robson J. Silva**, [2]**Savia Gavazza**, [2]**Lourdinha Florencio**, [2]**Mario T. Kato**

[1] Federal Institute of Pernambuco, Department of Infrastructure and Civil Construction, Av. Prof. Luiz Freire 500, Cidade Universitária, Recife PE, Brazil, (julianamorais@recife.ifpe.edu.br)

[2] Federal University of Pernambuco, Department of Civil Engineering, Laboratory of Environmental Sanitation, Av. Acadêmico Hélio Ramos s/n, Cidade Universitária, Recife PE, Brazil

[3] Federal University of Pernambuco, Academic Center of Agreste, Laboratory of Environmental Engineering, Rodovia BR-104, Km 62, Nova Caruaru, Caruaru PE, Brazil

Keywords: Activated sludge, domestic sewage, ponds, stone filters, UASB reactor

Abstract

This study of reliability analysis was based on the effluent data of four wastewater treatment plants (WWTPs) with different treatment technologies. The effluent constituents considered were: total COD, filtered COD, TSS, N-NH$_4$ and *Escherichia coli.* The centralized Janga WWTP (activated sludge) obtained reliability levels of 85, 80, 90 and 70% for total COD, filtered COD, TSS and N-NH$_4$, respectively. The decentralized Mangueira WWTP (UASB reactor + polishing pond) presented reliability levels of 40, 80, 40 and 85% for total COD, filtered COD, TSS and N-NH$_4$, respectively. In the decentralized Rio Formoso WWTP (UASB reactor + polishing pond + stone filter), it was observed an increase in the level of reliability which was attributed to the stone filters; the reliability levels were 70, 90, 85 and 85% for total COD, filtered COD, TSS and N-NH$_4$, respectively. Additionally, there was an excellent *E. coli* reliability level of 99%. The decentralized Petrolândia WWTP (facultative + 2 maturations ponds) obtained 60, 70, 70 and 30% for total COD, filtered COD, N-NH$_4$ and *E. coli* reliability level, respectively. The contribution of this reliability study consisted in generating data, which can be used by designer/operators in the performance evaluation of other WWTPs, considering the quality of the effluent.

1 Introduction

The concept of reliability in the design and operation of wastewater treatment plants, which involves the statistical analysis of the performance of treatment systems, has been applied and recommended by several authors for the evaluation of the treatment process [1, 2]. The reliability of a system can be defined as the probability of achieving adequate performance for a specific period of time under particular conditions. Due to the quality variations, wastewater treatment plants should be designed to produce effluents with average concentrations lower than those of the discharge standards [3, 4] developed a coefficient of reliability (COR) that relates the mean constituent values (MCP) to the standard that should be achieved based on probabilistic analysis. The ability to estimate treatment reliability enhances the risk manager's arsenal and can be used in various circumstances, such as providing the staff with the ability to demonstrate whether incorporating an alternative technology into a process train would provide the same degree of public health protection as a conventional or accepted technology [5]. The decentralized WWTP could be a cheaper alternative for developing countries, especially because of the warm or hot climate. Centralized WWTP should be used for large population with cautions due to the high consumption of electrical energy and more complex operation. In developing countries, the WWTPs have been shown operational problems that are not yet resolved. These limitations need to be studied and corrected aiming to not impair the effluent quality. The objective of this article is to present the performance results and statistical analyses of reliability obtained in centralized and decentralized WWTPs. The study was based on the effluent data of four plants with different treatment technologies.

2 Materials and Methods

2.1 Wastewater Treatment Plants

The field work was conducted at four WWTPs. The treatment systems, type of operation, scale, served population, and flow are shown in Table 1.

Table 1: Main characteristics of the wastewater treatment plants

WWTP	Operation	Scale	Treatment system	Population (inhab)	Flow (L/s)
Mangueira	Decentral	Medium	UASB reactor + polishing pond	18,000	32
Petrolândia	Decentral	Small	Facultative + 2 maturation ponds	8441	17
Janga	Central	Large	Activated sludge	451,900	300
Rio Formoso	Decentral	Medium	UASB reactor + polishing pond + stone filters	22,060	40

UASB = upflow anaerobic sludge bed

The monitoring program involved physical-chemical and microbiological analyses in the influent and effluent of the treatment units based on APHA [6].

2.2 Reliability analysis

Coefficients of Reliability (COR) has been applied and recommended by several authors for the assessment of treatment processes [1-3, 7]. The first step involves determining the coefficient of variation of the constituents and calculating the standardized normal variant ($Z_{\alpha-1}$) for reliability levels.

$$Z_{\alpha-1} = \frac{\ln X_s - \left[\ln \mu_x - \frac{1}{2}\ln(1 + CV_x^2)\right]}{\sqrt{\ln(1 + CV_x^2}}$$

Where: $Z_{\alpha-1}$ = standardized normal variant; X_s = effluent concentration as specified by the discharge standard (mg/L); μ_x = average distribution; CV = coefficient of variation.

In the following step, the reliability coefficient (COR) is calculated in terms of compliance of effluent with discharge standard, using the equation below:

$$\text{COR} = \sqrt{CV^2 + 1} \cdot \exp\left[-Z_{1-\alpha} \sqrt{\ln(CV^2 + 1)}\right]$$

With the coefficients of reliability obtained, it is possible to determine the design/operation values required for the effluent to meet the discharge standards.

$$MCP = (COR). Xs$$

Where: MCP = mean concentration of project (design/operational values) (mg/L).

The effluent constituents and the goals adopted for the present study were the following: total and filtered chemical oxygen demand (COD, 90 mg/L), total suspended solids (TSS, 60 mg/L), ammonium nitrogen (N-NH$_4$, 20 mg/L) and *Escherichia coli* (10^3 MPN/100 mL). These values are in agreement with legislation [8-10]. The coefficient of reliability of the components of interest was calculated for 80, 85, 90, 95, and 99% of reliability levels. The effluent constituents selected for the reliability analysis of the plants were as follows: (a) Mangueira: total and filtered COD and TSS; (b) Petrolândia: total and filtered COD, TSS, N-NH$_4$ and *E. coli.*; (c) Janga: total and filtered COD, TSS and N-NH$_4$; (d) Rio Formoso: total and filtered COD, TSS, N-NH$_4$ and *E. coli.*

3 Results and Discussion

3.1 Performance analysis of the wastewater treatment plants

In general, a great variability was noticed in the effluent concentrations and in the removal efficiencies, considering the analyzed constituents and the treatment technologies. Influent concentrations from urban WWTPs are inherently variable due the environmental characteristics, such as: temperature, pH, DO, alkalinity, COD and NTK concentrations. The effluent variability depends on the behavior of the influent loads and the treatment process. The results of the parameters obtained from the Mangueira, Petrolândia, Janga and Rio Formoso plants are summarized in Table 2; the treatment efficiencies are shown in Figure 1.

Table 2: Results of the main parameters monitored for the four plants

Parameters	Wastewater Treatment Plants			
	Mangueira	Petrolândia	Janga	Rio Formoso
	UASB + polishing pond	Facultative + 2 maturation ponds	Activated sludge	UASB reactor + polishing pond + stone filter
Temperature $_{in}$	$29.0 \pm 1.5^{(29)}$	$28.3 \pm 1.7^{(6)}$	$28.0 \pm 2.8^{(7)}$	$27.1 \pm 2.0^{(15)}$
pH $_{in}$	$7.2 \pm 0.2^{(60)}$	$7.4 \pm 0.4^{(6)}$	$7.0 \pm 0.3^{(7)}$	$6.9 \pm 0.4^{(14)}$
DO $_{ef}$	$7.0 \pm 2.7^{(44)}$	$6.4 \pm 2.9^{(8)}$	$2.6 \pm 0.6^{(7)}$	$2.8 \pm 1.4^{(16)}$
Total alkalinity $_{in}$	$222 \pm 24^{(56)}$	$204 \pm 26^{(9)}$	$281 \pm 20^{(7)}$	$155 \pm 46^{(13)}$
Total alkalinity $_{ef}$	$206 \pm 24^{(56)}$	$183 \pm 15^{(10)}$	$251 \pm 42^{(7)}$	$146 \pm 55^{(14)}$
Total solid $_{in}$	$632 \pm 140^{(31)}$	-	$514 \pm 191^{(7)}$	$689 \pm 1014^{(12)}$
Total volatile solid $_{in}$	$229 \pm 103^{(31)}$	-	$195 \pm 81^{(7)}$	$316 \pm 288^{(12)}$
Total susp. solid $_{in}$	$168 \pm 365^{(30)}$	-	$62 \pm 18^{(7)}$	$101 \pm 59^{(11)}$
Volatile susp. solid $_{in}$	$102 \pm 380^{(29)}$	-	$49 \pm 23^{(7)}$	$146 \pm 55^{(14)}$
Total solid $_{ef}$	$585 \pm 102^{(26)}$	-	$449 \pm 123^{(7)}$	$905 \pm 861^{(13)}$
Total volatile solid $_{ef}$	$187 \pm 76^{(26)}$	-	$102 \pm 68^{(7)}$	$179 \pm 876^{(13)}$
Total susp. solid $_{ef}$	$88 \pm 120^{(26)}$	-	$32 \pm 7^{(7)}$	$34 \pm 28^{(13)}$
Volatile susp. solid $_{ef}$	$68 \pm 85^{(26)}$	-	$15 \pm 12^{(7)}$	$16 \pm 14^{(13)}$
COD $_{in\ tot}$	$395 \pm 149^{(53)}$	$444 \pm 251^{(10)}$	$520 \pm 62^{(7)}$	$384 \pm 132^{(13)}$
COD $_{in\ fil}$	$151 \pm 71^{(51)}$	$148 \pm 100^{(10)}$	$197 \pm 19^{(7)}$	$76 \pm 34^{(12)}$
COD $_{in\ fil}$/COD $_{in\ tot}$	$0.38 \pm 0.11^{(46)}$	$0.33 \pm 0.18^{(10)}$	$0.21 \pm 0.03^{(12)}$	$0.21 \pm 0.03^{(12)}$
COD $_{ef\ tot}$	$189 \pm 123^{(47)}$	$136 \pm 50^{(10)}$	$55 \pm 28^{(7)}$	$82 \pm 78^{(15)}$
COD $_{ef\ fil}$	$64 \pm 51^{(45)}$	$72 \pm 36^{(10)}$	$47 \pm 16^{(7)}$	$33 \pm 21^{(15)}$
N- TKN $_{in}$	$36 \pm 8^{(35)}$	$50 \pm 14^{(10)}$	$33 \pm 2^{(7)}$	$26 \pm 14^{(8)}$
N-TKN $_{ef}$	$28 \pm 4^{(32)}$	$32 \pm 4^{(11)}$	$22 \pm 7^{(7)}$	$17 \pm 5^{(9)}$
N-NH$_4^+$ $_{in}$	$26 \pm 8^{(31)}$	$30 \pm 8^{(10)}$	$26 \pm 2^{(7)}$	$21 \pm 11^{(10)}$
N-NH$_4^+$ $_{ef}$	$19 \pm 3^{(34)}$	$21 \pm 3^{(11)}$	$24 \pm 5^{(7)}$	$12 \pm 3^{(11)}$
N-NO$_2^-$ $_{in}$	-	$0.01 \pm 0.01^{(5)}$	-	$0.1 \pm 0.3^{(8)}$
N-NO$_2^-$ $_{ef}$	-	$0.01 \pm 0.02^{(5)}$	$6 \pm 8^{(7)}$	$0.2 \pm 1.3^{(10)}$
N-NO$_3^-$ $_{in}$	-	$4.4 \pm 5.1^{(3)}$	-	$0.04 \pm 0.40^{(11)}$
N-NO$_3^-$ $_{ef}$	-	$5.7 \pm 5.2^{(2)}$	$1.0 \pm 0.1^{(7)}$	$0.1 \pm 1.4^{(8)}$
Phosphorus $_{in}$	$3.0 \pm 0.9^{(16)}$	$5.8 \pm 1.8^{(7)}$	$2.0 \pm 1.0^{(7)}$	$1.3 \pm 1.0^{(8)}$
Phosphorus $_{ef}$	$3.0 \pm 0.6^{(16)}$	$6.0 \pm 1.0^{(7)}$	$4.0 \pm 1.0^{(7)}$	$2.4 \pm 0.7^{(10)}$
E. coli $_{in}$	$3.22 .10^6 \pm 1.48 .10^{8\ (3)}$	$1.85.10^6 \pm 3.25.10^{6\ (4)}$	-	$1.17.10^7 \pm 2.52.10^{7\ (6)}$
E. coli $_{ef}$	$6.14 .10^5 \pm 1.47 .10^{8\ (3)}$	$2.36.10^4 \pm 2.23.10^{5\ (4)}$	-	$2.00.10^3 \pm 1.19.10^{3\ (7)}$

Values are geometric averages ± standard deviation; in parentheses - number of samples; concentrations are in mg/L, except for temperature (°C) and pH; in = influent; ef = final effluent; tot = total; fil = filtrate; DO = dissolved oxygen; TKN = total Kjeldahl nitrogen.

3.1.1 Decentralized Mangueira WWTP (UASB + polishing pond)

The average total influent COD was of low-strength with concentrations of 395 ± 149 mg/L. The Mangueira WWTP performed well in terms of organic material removal with the average effluent total and filtered COD of 189 ± 123 and 64 ± 51 mg/L, respectively (Table 2). The total and filtered COD removal efficiency of the UASB reactor was 52±13 and 75±12%, respectively; and 52±18, 79±13%, respectively, for the entire plant (Figure 1a). The filtered COD removal efficiency for the polishing pond did not include the algal mass in the effluent as organic matter. The UASB showed poor TSS removal efficiency with an average value of 43 ± 18%. The same occurred with the final TSS of the plant since the average of only 37 ± 31% was strongly influenced by the algal biomass of the polishing pond effluent. The $COD_{tot-fil}$, $COD_{tot-tot}$ and TSS efficiencies of the UASB were 81, 60 and 43%, respectively. The poor UASB efficiencies for $COD_{tot-tot}$ and TSS were attributed to the loss of solids in the effluent because of a lack of routine discharge of excess sludge and some other deficiencies that were observed in the operation. Nevertheless, the global COD removal of the plant was 79% because of the good efficiency of the reactor. The favorable climatic conditions contributed greatly to the efficient removal of the organic material.

3.1.2 Decentralized Petrolândia WWTP (Facultative + 2 maturation ponds)

The average total COD removal efficiency in the Petrolândia WWTP was 65 ± 14%. However, when considering the filtered COD and not taking into account the biomass algal of the polishing pond effluent, the efficiency increases to 80 ± 9%, with the average effluent total and filtered COD concentrations of 136 ± 50 and 72 ± 36 mg/L, respectively (Table 2). The *E. coli* removal efficiency in the facultative and maturation ponds was 98.9 ± 1.4% (Figure 1b), with a final effluent of 2.36 x 10^4 (Table 2). The favorable climatic conditions (high temperature and insolation) contributed significantly to that relatively high efficiency removal of organic material and *E. coli*. The municipality is located in the semi-arid region of the Brazilian Northeast with average and maximum temperatures of 28 and 31 °C, respectively. On the other hand, the presence of two potentially toxic species of *Cyanophyta, Oscillatoria* sp. and *Microcystis aeruginosa* was observed, which indicates that caution is required when considering the final destination of the treated effluent [11].

3.1.3 Centralized Janga WWTP (activated sludge)

The Janga WWTP presented excellent results in terms of organic matter removal, with total and filtered COD efficiency of 88±6 and 89±3%, respectively (Figure 1c). However, the nitrogen removal was inefficient with values of only 30±25%. The effluent concentration of ammonium nitrogen was 24±5 mg/L (Table 2). In the last years, the plant has presented several operational problems like irregular return sludge flow, waste sludge flow and aeration capacity. The growth of the nitrifying autotrophics may have been affected by the low DO availability, inhibiting the nitrification process. The centralized Janga WWTP serves a population of 451,900 inhabitants and has been under operation of 35 years. Today, the plant is under several construction and operational improvements.

3.1.4 Decentralized Rio Formoso WWTP (UASB reactor + polishing pond + stone filters)

The Rio Formoso UASB showed average efficiencies of 32±19 and 79±7 % for total and filtered COD removal, respectively (Figure 1d). The poor TSS removal with efficiency of 30±40% undermined the best performance of the anaerobic reactor. The average effluent total COD, filtered COD and TSS were 82±78, 33±21 and 101±59 mg/L, respectively (Table 2). The nitrogen removal was inefficient, with an average value of only 50%. The *E. coli* removal was highly significant, with an average removal of 99.98% and effluent concentrations of 2.00×10^3. This excellent removal is associated with the operation of the polishing pond and the stone filters. The stone filters are also effective in removing cyanobacteria. The specific genera *mcyE* was not found, indicating the absence of toxins produced by *Microcystis sp.* [12].

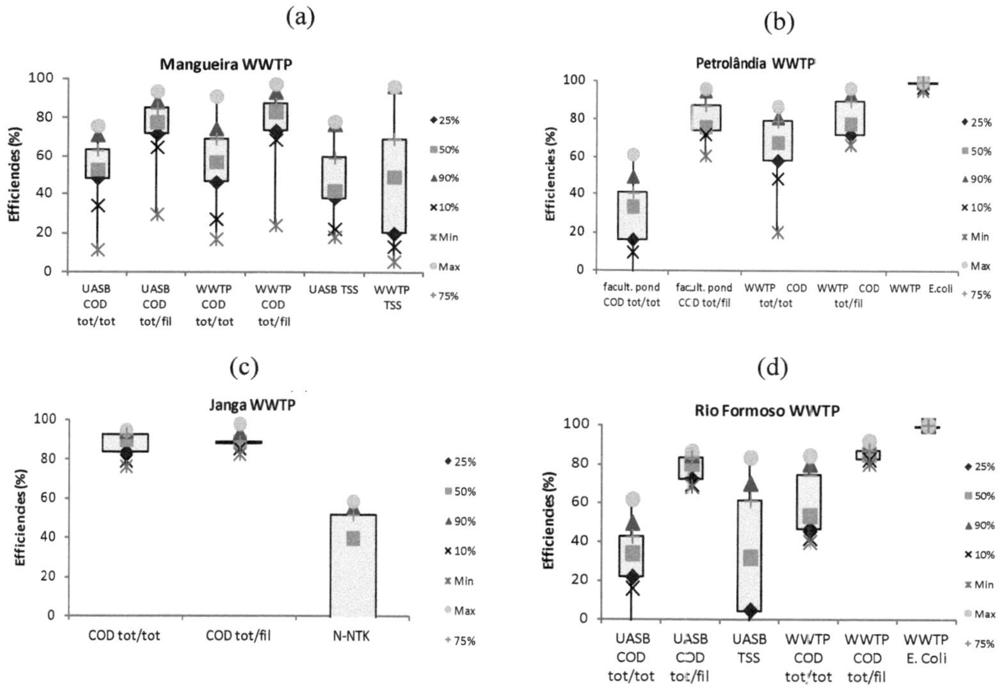

Figure 1: Treatment efficiency of the plants: (a) Mangueira (decentralized); (b) Petrolândia (decentralized); (c) Janga (centralized); (d) Rio Formoso (decentralized), where $COD_{tot-tot}$ is the removal efficiency based on total influent and effluent COD, and $COD_{tot-fil}$ is based on the total influent COD and filtrate effluent COD.

3.2 Coefficient of reliability and mean concentration for the discharge standards or treatment targets

The coefficient of reliability of the components of interest was calculated for 80, 85, 90, 95, and 99% of reliability levels. Table 3 shows the COR values of the coefficient of reliability, the mean concentration of the project (MCP), the mean concentration obtained (MCO), and reliability level obtained (RLO) for the selected effluent constituents in the wastewater treatment plants. The interpretation of the reliability parameters is obtained from the Table 3. For example, for a 95% level of reliability, a COR of 0.30 for the effluent COD_{tot} of the Mangueira WWTP was obtained. This means that, in order to comply with the adopted standard in 95% of the operating time, the mean concentration of project (MCP) should be 27 mg/L. In practice, the mean concentration obtained were 189 mg/L, much higher than the concentrations of project. When the parameters did not achieve 80% of reliability level, the RLO was calculated from the coefficient of variation, the standard normal variant ($Z_{\alpha-1}$) and COR were obtained.

The analysis of the centralized and decentralized WWTP showed great variability in the coefficients of variation and reliability. For high levels of reliability, the coefficient of variation varied inversely with the values of the COR. Low coefficient of variations and consequently high COD values do not necessarily imply a good performance, but simply a more stable operating condition [3]. It is important to emphasize that the reliability analysis does not question the potentiality of treatment technology, but emphasizes that the operational instability in a WWTP may significantly impair any treatment process. This fact implies that more accurate and stable operation is required, without allowing large variations in the effluent concentrations of the treatment units.

The centralized Janga WWTP (activated sludge) obtained the lowest COR values, with high reliability level of 85, 80, 90 and 70% for total COD, filtered COD, TSS and $N-NH_4$, respectively (Table 3). Janga WWTP achieved very good performance in terms of organic matter and TSS, with a reliability level above 80%, indicating that the results reached the adopted goal in 80% of the operating time. The best reliability level was found for activated sludge when analyzed 166 WWTPs in Brazil [13]. The aeration capacity in the Janga plant was the main operational problems that inhibited the nitrification process and consequently, affected the level of reliability of this parameter.

The decentralized Mangueira WWTP (UASB reactor + polishing pond) presented reliability levels of 40, 80, 40 and 85% for total COD, filtered COD, TSS and $N-NH_4$, respectively (Table 3). The filtered COD and $N-NH_4$ presented good reliability levels, reaching the adopted goal in 80% of the operating time. The total COD and TSS reliability levels were influenced by algal mass in the polishing pond effluent.

Table 3: Coefficient of reliability (COR), mean concentration of project (MCP), mean concentration obtained (MCO) and reliability level obtained (RLO).

	Coefficient of reliability (COR)									
Reliability level (%)	Mangueira WWTP					Petrolândia WWTP				
	COD_{tot}	COD_{fil}	TSS	N-NH_4	*E.Coli*	COD_{tot}	COD_{fil}	TSS	N-NH_4	*E. coli*
80	0.72	0.71	0.72	1.31	-	0.81	0.77	-	1.28	0.70
85	0.59	0.59	0.63	0.89	-	0.62	0.60	-	0.88	0.59
90	0.45	0.47	0.54	0.55	-	0.44	0.44	-	0.55	0.47
95	0.30	0.33	0.43	0.27	-	0.27	0.28	-	0.27	0.42
99	0.14	0.17	0.28	0.07	-	0.10	0.12	-	0.07	0.19

	Coefficient of reliability (COR)									
Reliability level (%)	Janga WWTP					Rio Formoso WWTP				
	COD_{tot}	COD_{fil}	TSS	N-NH_4	*E.Coli*	COD_{tot}	COD_{fil}	TSS	N-NH_4	*E. coli*
80	0.88	0.81	1.10	1.09	-	0.71	0.75	0.70	0.99	0.70
85	0.64	0.62	0.78	0.77	-	0.58	0.59	0.59	0.71	0.61
90	0.45	0.44	0.50	0.50	-	0.46	0.44	0.47	0.47	0.50
95	0.26	0.27	0.26	0.26	-	0.32	0.29	0.34	0.26	0.38
99	0.10	0.10	0.08	0.08	-	0.16	0.13	0.19	0.08	0.22

	Mean concentration of project (MCP)									
Reliability level (%)	Mangueira WWTP					Petrolândia WWTP				
	COD_{tot}	COD_{fil}	TSS	N-NH_4	*E.Coli*	COD_{tot}	COD_{fil}	TSS	N-NH_4	*E. coli*
80	65	63	43	26	-	73	69	-	26	$6.32.10^3$
85	53	53	38	18	-	56	54	-	18	$5.32.10^3$
90	40	42	33	11	-	40	40	-	11	$4.27.10^3$
95	27	30	26	5	-	24	25	-	5	$3.09.10^3$
99	13	16	17	1	-	9	11	-	1	$1.69.10^3$
MCO	189	64	88	19	-	136	72	-	30	$2.36.10^4$
RLO (%)	40	80	40	85	-	60	70	-	70	30

	Mean concentration of project (MCP)									
Reliability level (%)	Janga WWTP					Rio Formoso WWTP				
	COD_{tot}	COD_{fil}	TSS	N-NH_4	*E.Coli*	COD_{tot}	COD_{fil}	TSS	N-NH_4	*E. coli*
80	76	73	66	22	-	64	67	42	20	$6.33.10^3$
85	57	56	47	15	-	53	53	35	14	$5.45.10^3$
90	40	40	30	10	-	41	40	28	9	$4.51.10^3$
95	24	24	16	5	-	29	26	21	5	$3.41.10^3$
99	9	9	5	2	-	15	11	11	2	$2.02.10^3$
MCO	55	47	32	24	-	82	33	34	12	$2.00.10^3$
RLO (%)	85	80	90	70	-	70	90	85	85	99

Units: mg/L, except for COR (a dimensional number), RLO (%) and *E. coli* (MPN/100 mL).

In the decentralized Rio Formoso WWTP (UASB reactor + polishing pond + stone filter), it was observed an increase in the level of reliability due to the stone filters, especially when compared with Mangueira WWTP (UASB reactor + polishing pond). The Rio Formoso reliability levels were 70, 90, 85, and 85% for total COD, filtered COD, TSS, $N-NH_4$, reaching the best performance and reliability between the three decentralized WWTPs. The *E. coli* parameter reached the adopted goal in 99% of the operating time with mean concentration of operation of 2.00×10^3. This excellent reliability level is associated with the operation of the polishing pond and the stone filters.

The decentralized Petrolândia WWTP (facultative + 2 maturations ponds) obtained 60, 70, 70 and 30% for total COD, filtered COD, $N-NH_4$ and *E. coli* reliability level, respectively. Despite the high *E. coli* removal efficiency (98.4±1.4%), the reliability level was low (30%) due to average effluent concentration of 2.36×10^4, which is above the goal adopted of 1.00×10^3 MPN/100 mL.

4 Conclusions

In terms of reliability analysis, the coefficients of variation and reliability showed great variability concerning the four WWTPs studied. Under the conditions of operation found in the WWTPs, a 95% of reliability level was hardly reached, but it ranged from 70 to 90% for filtered COD, and it was higher than 85% for TSS, except for Mangueira WWTP. The observed poor performance in terms of nitrogen removal was expected, since none of the technologies studied has been designed for nutrient removal. Ammonification or partial nitrification was obtained in the WWTPs. The results indicate that the four plants need operational improvement in order to meet the Brazilian discharge standards. The recommendation is that monitoring should be conducted on an individual basis from the influent characterization, loads applied to environmental conditions. The operating parameters are of probabilistic nature with variable intensity and concentrations.

Nevertheless, this reliability study consisting of generating data, can contribute to designers and sanitation companies in the performance evaluation of the WWTPs, considering the quality of the effluent. It is also important to emphasize that the obtained results showed the existing reality of four plants, but not their treatment technologies, as well as that the WWTPs are able to achieve better performance and reliability levels.

5 Acknowledgements

The authors are grateful to the Brazilian agencies, the National Council for Scientific and Technological Development (CNPq) and the National Financing Agency of Studies and Projects (FINEP), both of the Ministry of Science, Technology and Innovation, for their support through the RENTED Project. DAAD and EXCEED Swindon Project are also acknowledged for the opportunity to participate at this expert workshop.

6 References

[1]   Naval, L.P., Wanderley, T.F., Evaluation of a reactor (UASB) startup without inoculum using physicochemical parameters and applying probabilistic model. In: Proceedings of the 27th Interamerican Congress on Sanitary and Environmental Engineering. , ABES, Rio de Janeiro, Brazil. 2000,

[2]   Metcalf & Eddy., Wastewater engineering: treatment, and reuse. 4th Ed. Metcalf & Eddy, Inc., New York, pp.1819, 2003.

[3]   Oliveira, S., von Sperling, M., Reliability analysis of wastewater treatment plants, Water Res. 2008, 42, 1182-1194.

[4]   Niku, S., Schroeder, E.D., Samaniego, F.J. Performance of activated sludge process and reliability-based design, J. Water Pollut. Control Fed., 1979, 51(12), 2841-2857.

[5]   Eisenberg, D., Soller,J., Sakaji;R., Olivieri, A., A methodology to evaluate water and wastewater treatment plant reliability. Water Sci. Technol. 2001, 43(10), 91–99.

[6]   APHA/AWWA/WEF, Standard Methods for the Examination of Water and Wastewater, 22nd Ed.: American Public Health Association, American Water Works Association, Water Environment Federation. Washington DC, 2012.

[7]   Morais, J.C., Kato, M., Florencio, L., Gavazza, S., Assessing the efficiency and operational problems of an anaerobic wastewater treatment plant over 13 years of monitoring. In: Proceedings of the X Latin American Workshop and Sympcsium on Anaerobic Digestion. IWA, Ouro preto - Minas Gerais, Brazil, 2011

[8]   MINAS GERAIS. Deliberação Normativa nº 10 de 16 de dezembro de 1986. Estabelece normas e padrões para qualidade das águas, lançamento de efluentes nas coleções de águas, e dá outras providências. Belo Horizonte: Conselho ce Política Ambiental de Minas Gerais – COPAM, 1986.

[9]   CONAMA. Resolução nº 430, de 13 de maio de 2011. Dispõe sobre as condições e padrões de lançamento de efluentes, complementa e altera a Resolução no 357, de 17 de março de 2005, do Conselho Nacional do Meio Ambiente - CONAMA. Disponível em http://www.mma.gov.br/port/conama/legiabre.cfm?codlegi=646, (accessed on 22/06/2011)

[10] WHO. Health guidelines for the use of wastewater in agriculture and aquaculture. Technical Report Series, 778. Geneva, Switzerland, 1989.

[11] Pastich, E. A., Gavazza, S., Casé, M. C. C., Florencio, L., Kato, M. T., Structure and dynamics of the phytoplankton community within a maturation pond in a semiarid region. Brazilian J. Biology, 2016, 76(1), 144-153.

[12] Santos, M. V. A., Kochling, T., Gavazza, S., Kato, M. T., Florencio, L., Horizontal flow rock filters used in the post-treatment of stabilization ponds to remove cyanobacteria and suspended solids. In: Proceedings of the Workshop on Water Pollution Control and Biomass Energy, Xi'an, China, 2015, 1. 129-139.

[13] Oliveira, S., von Sperling, M. Performance evaluation of UASB reactor systems with and without post-treatment. Water Sci. Technol.. 2009, 59, 1299-1306

SINGLE AND TWO PHASE THERMOPHILIC ANAEROBIC DIGESTION OF WASTE ACTIVATED SLUDGE: PERFORMANCE AND ENERGY ASSESSMENT

[1,3]**Wanderli Leite**, [2]**Marco Gottardo**, [2]**Paolo Pavan**, [1]**Paulo Belli Filho**, [3]**David Bolzonella**

[1]Department of Sanitary and Environmental Engineering, Federal University of Santa Catarina, Trindade, Florianópolis, Brazil. Email addresses: wanderli.leite@posgrad.ufsc.br, paulo.belli@ufsc.br
[2]Department of Informatics, Statistic and Environmental Science, University Ca Foscari of Venice, Venice, Italy. marco.gottardo@unive.it, pavan@unive.it
[3]Department of Biotechnology, University of Verona, Strada Le Grazie 15. 37134 Verona, Italy. david.bolzonella@univr.it

Keywords: Anaerobic digestion, single-phase reactor, thermophilic, two-phase reactor, waste activated sludge

Abstract

Single and 2-phase anaerobic digesters were conducted at the same conditions of organic loading and retention time in order to compare their performances on sludge stabilizing and biogas yields. There was a slight increase in terms of volatile solids' removal, rising from 34% in the single-phase system to 38% in the 2-phase system. As a consequence, its global specific biogas production was greater around 0.31 m^3/kgTVS$_{fed}$d. The 2-phase system produced 15% more energy than the single stage system. Furthermore, the positive energy balance developed for both systems depicted none costs with external energy for sludge heating purposes. The estimated payback period to recovery the investment of the realization of the hydrolytic reactor was about 3 years.

1 Introduction

Adequate management of sludge is a challenge for the majority of WWTPs due to the cost involved, which reach approximately 40% of all operational input [1]. Therefore, treatment and disposal of this residual sludge is receiving increasing attention. Anaerobic digestion (AD) of waste activated sludge (WAS) is widely considered the most important and appropriate method for sludge treatment before its final disposal. The AD of WAS enhances the value of this organic residue, while the produced biogas can cover part of the energy requirements of the activated sludge process [2,3].

It is reported that the initial hydrolysis of particulate organic matter to soluble substances is the rate-limiting step of anaerobic digestion of WAS [4,5]. The phase separation of hydrolysis/ fermentation from methanogenesis in different reaction environments may lead to a larger biogas yield and optimize the overall reaction rate, with regard to stability and substrate degradation efficiencies in both reactors [6,7]. When applying thermophilic or extreme thermophilic processes, the energy balance of the system should be analyzed in order to verify the possibility to guarantee at least self-sustaining WAS digestion [8, 9]. Additionally, operating variables of digesters including hydraulic retention time (HRT), temperature, and the influent solids concentration influence dewatering of the digestate, so that, carefully-controlled laboratory tests are necessary for their determination [10].

This paper presents the results of a pilot scale study where both single phase and two-phase anaerobic digestion configurations for stabilizing WAS were tested.

2 Materials and methods

2.1 Substrate and inoculum

The substrate used in this study consisted of municipal WAS originated from a 70,000 PE wastewater treatment plant that treats 19,000 m^3/d of municipal wastewater adopting the BNR (Johannesburg scheme) process. The plant is located in Treviso (northern Italy). Average characteristics of WAS after gravity thickening were those shown in Table 1.

Table 1: Feed characteristics

Variable	Unit	Average ± Standard Deviation
pH	-	6.7 ± 0.2
TS	g/kg	62.4 ± 12.4
TVS	%	68.2 ± 3.0
COD	g/kg	48.2 ± 7.9
TKN	g/kg	3.14 ± 0.71
TP	g/kg	0.90 ± 0.40

The single-phase reactor was seeded using the anaerobic sludge originated from a 2,200 m^3 anaerobic digester located also in this plant, which treats WAS and separately collected biowaste at a working temperature of 35 °C. Biomass acclimation to thermophilic conditions occurred towards a single step temperature conversion as suggested by Cecchi et al. [11].

2.2 Experimental setup

AD tests were carried out at pilot-scale in thermophilic conditions and consisted of a first AD trial (run 1) in single phase and a second AD trial, in which a second reactor (fermenter) was added for the pre-treatment and enhanced hydrolysis of the WAS before its treatment in the thermophilic methanogenic digester. The pilot-scale anaerobic digester (methanogenic phase) operated as the single-phase reactor in run 1 and as fermenter phase during run 2. It was a 0.15 m^3 continuous stirred reactor (CSTR) of working volume. A 0.23 m^3 CSTR reactor was used as the methanogenic phase during run 2. Both reactors had an external water jacket for temperature control through a PT-100 based thermostat. Regarding the operational conditions, the methanogenic reactor operated with a HRT of 20 days in run 1 and 18 days in run 2, and an OLR of 2.2 kgTVS per m^3 of reactor per day, while the fermenter reactor used in run 2 operated with a HRT of 2 days and a OLR of 15 kgTVS per m^3 of reactor per day.

2.3 Analytical methods

Samples were collected three times a week from all reactors to monitor stability parameters as pH, volatile fatty acids (VFAs) content and speciation, total and partial alkalinity, ammonia, total solids (TS), and volatile solids (TVS). Once a week, substrate and effluents were monitored in terms of COD, total nitrogen (TKN), and total phosphorus (TP) on dried samples. All analyses were carried out in accordance to the Standard Methods [12], while VFA were determined by GC in a Carlo Erba gas-chromatograph (GC) equipped with a flame ionization detector (FID) (200 $^\circ$C).

Daily produced biogas was measured with two flow meters (Ritter Company, drum-type wet-test volumetric gas meters), fitted on the reactors. Biogas composition was determined by a GC equipped with a HP-Molesieve column (30 m x 0.3 mm x 0.25 μm film thickness) employing thermal conductivity detection (TCD). Gas composition was also defined using a portable infrared gas analyzer (GA 2000, Geotechnical Instruments).

Dewatering tests were conducted using a capillary suction time (CST) instrument (Model 319, Trition, UK) [12] and the specific resistance to filtration test (SRF) according to IRSA-CNR [13]. All tests were conducted in triplicate. Student´s t-tests were performed at 95% confidence level in order to analyze differences between means (Statistica software, StatSoft, USA).

2.4 Heat requirements and energy balances

In order to estimate heat balances for the AD systems, operating in a hypothetical WWTP with a size of 100,000 PE was considered. A specific heat request of 1 kcal/kg·°C, the sludge temperature of 10 $^\circ$C, a combustion heat for biogas of 5,500 kcal/m^3, thermal and electrical yields of the heat and power co-generation unit (CHP) of 50% and 40%, respectively, were used. To define the heating request for sludge flow the sludge production was set at 60 g dry matter per person equivalent per day. Total heat losses were estimated considering the dimensions of the reactors and the typical construction specifications. Specific yields were determined from the data obtained experimentally.

3 Results and discussion

3.1 Substrate characteristics

The average characteristics of the WAS used as feed to both single stage digester and the fermenter reactor of the 2-phase system are shown in Table 1. The COD:TVS ratio was 1.2 rather than a typically 1.4 and can be ascribed to the relatively poor F/M ratio in the activated sludge process which indicates that WAS was stabilized in the BNR process due to the high SRT in the process (15 days).

The most significant observation of this study was the relatively high solids content of waste activated sludge fed to the anaerobic digester: The average concentration of TS in the fed sludge was 6.2%. In this way, the volatile solids concentration was around 4.3%. As a consequence, the OLR applied for both methanogenic digesters was some 2.2 ± 0.5 kgTVS/m^3 per day and 14.5 ± 1.7 kgTVS/m^3 per day for the fermenter reactor.

3.2 Effluent characteristics

Table 2 shows the effluent characteristics, stability parameters, and yield of single stage and 2-phase AD systems. Ammonium concentrations had similar values for both digesters (around 1,400 mg NH_4^+/L, $p>0.05$). Even though free ammonia (FA) is toxic to methanogenic bacteria, no traces of ammonia toxicity were observed during the experimentation. Total and partial alkalinity were higher in the 2-phase digester (one-tailed t-test, p-value <0.05 for both cases), and this was supported also by the higher ammonification rate in this reactor (further discussed).

Table 2: Effluent characteristics, stability parameters and yields for both experiments

Average values	Single phase [1]	2-phase fermenter [2]	2-phase digester [2]
Effluent characteristics			
pH	8.2 ± 0.1 [a]	6.8 ± 0.1	8.2 ± 0.1 [e]
P-Alkalinity (mg $CaCO_3$/L)	$3,049\pm421$ [a]	655 ± 74	$3,603\pm222$ [e]
T-Alkalinity (mg $CaCO_3$/L)	$5,295\pm692$ [a]	$4,029\pm193$	$5,634\pm161$ [e]
Ammonia (mg NH_4^+/L)	$1,433\pm247$ [a]	$1,199\pm51$	$1,490\pm76$ [e]
Total VFA (mg COD/L)	$1,023\pm543$ [a]	$11,478\pm1431$	849 ± 128 [e]
TS (g/kg)	56.3 ± 8.4 [b]	52.4 ± 5.6	42.8 ± 1.2 [f]
TVS(g/kg)	28.5 ± 3.3 [b]	33.4 ± 5.7	26.5 ± 1.6 [f]
TP (g/kg TS)	16.1 ± 3.5	17.6 ± 0.5	21 ± 0.2
TKN (g/kg TS)	31.7 ± 9.7 [c]	37.8 ± 2.7	36.4 ± 0.7 [g]
COD (g O_2/kg TS)	605 ± 67 [c]	596 ± 136	639 ± 94 [g]
Soluble COD (mg O_2/L)	$5,169\pm1,176$ [c]	$10,758\pm382$	$5,413\pm805$
Yields			
Biogas (m^3/d)	0.07 ± 0.01 [d]	0.13 ± 0.02	0.13 ± 0.01 [h]
TVS reduction (%)	34 ± 3.5	22 ± 4.6	38 ± 2.7
CH_4 content (%) [d]	62 ± 5	35 ± 3	69 ± 2 [h]

[1]Single-phase valid N: [a]36; [b] 26; [c] 14; [d] 38; [2]2-phase valid N: [e] 16; [f] 19; [g] 4; [h] 21.

Total VFA:SCOD ratio was high in the fermentate during the 2-phase AD and revealed that 90% of all soluble COD were volatile fatty acids (determined as sum of the acids from C2 – acetic to C7 – heptanoic). The digesters, however, presented both similar and low values of VFA:SCOD ratio (p>0.05) since the anaerobic microorganisms were able to convert VFA at high efficiency.

According to the data presented in Table 2, the yields of the single stage anaerobic digester regarding biogas production rate, GPR, was in the range 0.5 ±0.1 $m^3_{biogas}/m^3_{reactor}$day, while the specific gas production, SGP per kg of volatile solids added to the reactor, were in the range 0.21±0.04 m^3_{biogas}. The average TVS removal was 34% instead of the typical 48-55% observed in digesters treating mixed sludges [14] Once the WAS was subjected to a pretreatment in the first reactor, where the volatile solids were hydrolyzed and metabolized to organic acids, the biogas production by methanogens in the 2-phase system was strongly facilitated. In fact, the specific methane production (SMP) was higher in the 2-phase digester (0.17±0.02 $m^3 CH_4$/ kgTVS$_{fed}$ versus 0.13±0.03 $m^3 CH_4$/kgTVS$_{fed}$ in the single-stage digester). Moreover, this digester presented GPR values of 0.55±0.05 $m^3_{biogas}/m^3_{reactor}$day, a SGP of 0.25±0.03$m^3_{biogas}$/kgTVS$_{fed}$ and a TVS removal efficiency of 38%.

Comparing the results of the single stage and two stage systems (Figure 1) it is evident from the data reported in Table 2 that there was a slight increase in terms of volatile solids removal, rising from 34%±3.5 to 38%±2.7, respectively. As a consequence, the average (global) specific biogas production for the 2-phase system was 0.31 m^3/kgTVS$_{fed}$.d, statistically different (p<0.01) and greater than the one observed in the single stage system (0.21 m^3/kgTVS$_{fed}$.d).

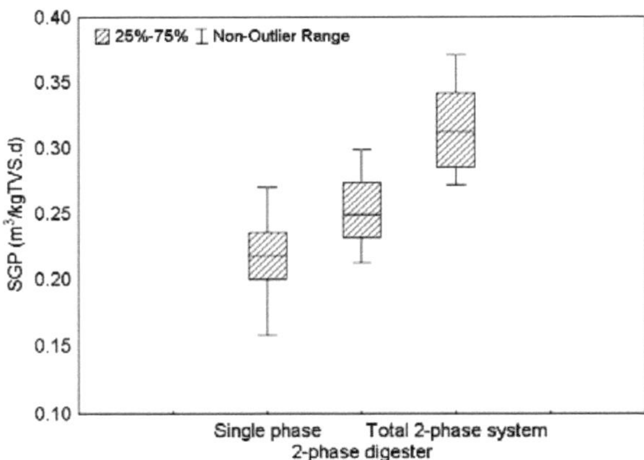

Figure 1: Box-plot representation of specific gas production for single and 2-phase systems

3.4 Energy assessment

The heat requirements and energy balance analyses were performed for a WWTP of 100,000 PE. A working volume of 2,000 m^3 was set for both single stage and 2-phase configuration. In the 2-phase system, volumes of 255 m^3 for the fermenter and 1,745 m^3 for the methanogenic reactor were considered. Figure 2 shows the energy balance for the single stage and 2-phase systems. Because of the enhanced biogas production there was a higher electrical energy production of 0.4 MWh/day for the 2-phase system. Further, it was observed that the 2-phase system produced 15% more total energy than the single stage system. In the meantime, energy losses from the 2-phase reactors walls were 14% greater than that in the single stage reactor because of the larger surface of the two reactors. However, despite the lower energy production for the single stage system and the higher heat losses for the 2-phase system, in all cases the energy balance was positive as confirmed by the difference between the potential heat energy production and the total energy requirements (residual heat energy). This result can be ascribed to the use of concentrated sludge (6% TS) as reported in previous studies [8].

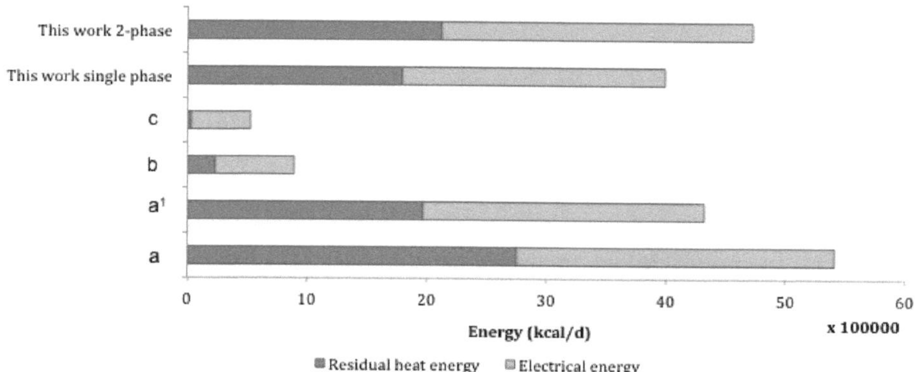

Figure 2: Energy balance for the single and 2-phase systems ([a][15], [c] [16], [b] [17]

Traditionally, mesophilic anaerobic digestion is more widely used compared to thermophilic systems, amongst other reasons, due to the energy demand and feasibility [18]. However, according to Figure 2, residual heat and electrical energy availability were often higher when compared with data from mesophilic digestion of WAS in literature. Indeed, mesophilic anaerobic sludge digestion can still achieve higher biogas productivity and then might represent an alternative to enhance the energy recovery from the sludge treatment when, for instance, a more efficient dynamic dewatering is intended to rise solids loading [19] or whenever sludge pretreatments are used to increase sludge biodegradability [20]. The main outcomes of the energy balance performed in this study and compared to other single phase anaerobic WAS cases are detailed in Table 3. As expected, the 2-phase system had the highest total heat consumption because of the larger heat losses from the walls of the reactors as described above.

Table 3: Main outcomes of energy balance carried out on the current single and 2-phase system and other AD studies.

RT[1]	GPR[2]	HFS[3]	HLD[4]	THC[5]	PP[6]	IRE[7,8]	Reference
Meso	0.60	1,852	578,055	579,907	3.09	178,265	[15] (a)
Thermo	0.53	3,333	954,501	957,835	2.73	157,043	[15] (a[1])
Meso	0.15	2,364	596,878	599,241	0.77	44,213	[17] (c)
Meso	0.11	1,667	578,055	579,722	0.57	32,717	[16] (b)
Thermo	0.50	4,500	954,501	959,001	2.56	147,375	This work 1-phase
Thermo	0.59	4,330	1.113,431	1.117,761	3.0	173,903	This work 2-phase[9]

[1] Temperature range (mesophilic or thermophilic); [2]Gas production rate ($m^3/m^3.d$); [3] Heat feeding sludge (kcal/d); [4] Heat losses digester (kcal/d); [5] Total heat consumption (kcal/d); [6]Power production (MWh/d); [7] Increased revenues from electricity (€/year); [8] Electricity cost adopted (€/MWh): 160 (no incentives).

3.4 Economics of the single and two-phase processes

In the previous scenario, revenues were 15% higher in the thermophilic 2-phase system which clearly can justify and support the preference of this reactor configuration regarding the production of renewable energy. Profits originated from current single stage system (147,375 €/year, Table 3) were similar to those reported by De la Rubia et al. [15] and higher than those achieved in mesophilic full scale system [17].

The payback period was defined as the time to recovery the amount invested to implement a hydrolytic reactor to improve sludge treatment, so a 2-phase AD system in the current WWTP (100,000 PE) would be performed. The single phase AD reactor was calculated based on data used in the energetic balance, i.e., 2000 m^3 of volume to treat an organic load of 4,500 kgTVS/d. Moreover, the hydrolytic reactor volume was set at 312 m^3 according to the operational HRT of 2.7 days and 17.3 days for the hydrolytic and methanogenic reactors, respectively. The investment for the hydrolytic reactor was determined in 234,000 Euros (750 Euros/m^3 reactor). Also the expenses for the adoption of necessary heat exchange, pumps and pipes were determined totaling 270,600 Euros [8]. Further, it was considered a dewatered excess digestate (25% on TS) as output from the sludge treatment line and the cost of 100 Euros per ton for sludge final disposal.

Savings obtained from the 2-phase system in respect to the single phase AD process were 95,451 Euros/year. About 35% of the profits were related to the increment of electric power sale (33,458 Euros/year) while the latest 65% came from the reduction on sludge final disposal (approximately 620 ton/year). Further, since energy balance in this scenario was also positive none costs with external energy were demanded for heating purposes. The payback period to recovery the investment of the realization of the hydrolytic reactor was about 3 years.

3.5 Mass balance for both single and 2-phase systems

According to data reported in Tables 1 and 2, the mass balances of the single main compounds were calculated. Figure 3 shows the mass balances of the 2 experimental trials for the parameter total and volatile solids, COD, total nitrogen, and total phosphorous.

The mass balances demonstrated that TS and TVS were only partially removed through biogas formation (22% and 28% for single phase and 2-phase systems, respectively) while the remaining fraction was removed in the digestate and also transformed into soluble organic compounds [8]. Regarding the COD, its conversion to biogas was similar to the TVS conversion and consistent with biogas production, ranging around 20% for both systems. The 2-phase system increased about 6% the organic solids conversion yield to biogas probably because submitting the WAS to a previous fermentation before the digester have favored the hydrolysis of solids that is well known as limiting step in the overall anaerobic digestion process, particularly regarding the high OLR applied in this reactor (15 kgTVS/m^3.d).

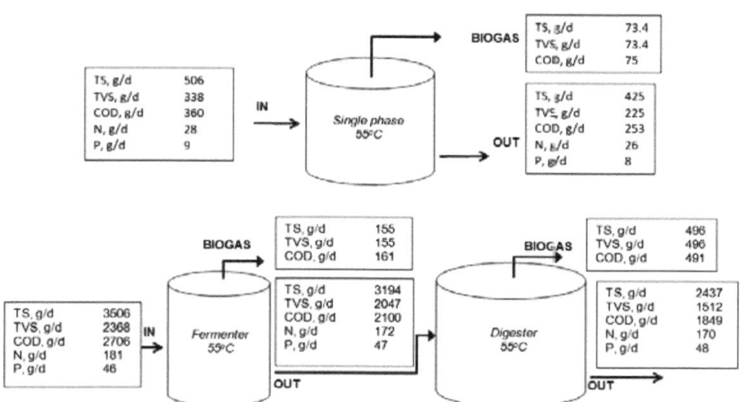

Figure 3: Mass balances for the single (above) and 2-phase (under) anaerobic systems

As for nutrients, nitrogen and phosphorous were just hydrolyzed: Nitrogen was almost present as organic nitrogen in the influent WAS while ammonia nitrogen accounted for 45% and 48% of total nitrogen in the single phase and 2-phase effluents, respectively.

3.8 Dewatering tests

Dewatering properties of the feed sludge and digestates, and optimal dose of cationic polymer conditioning (Hidrofloc C 675-Hydrodepur) were evaluated by CST and SRF tests. The polymer dosages performed were (measured in g/kgTS): 5.9, 6.75 and 8.8 for the single stage digestate and 5.0 and 8.7 for the 2-phase digestate (Table 4). The WAS used as feed was readily filtered without conditioning; in particular, its CST value was 139±6 s and SRF value was 1.64 x 10^8 m/kg (± 4.74 x10^7). On the other hand, performing dewatering tests with the single stage digestate, the higher the dosage of conditioning polymer the lower were CST and SRF results. Particularly, the greater reduction for the CST (37%) was obtained increasing the dosage from 5.9 g/kgTS to 6.75 g/kgTS.

The 2-phase digestate showed better filtration performances. Filtration tests were successfully performed even without polymer conditioning, and values of 1602±242 s and 1.51 x 10^9 m/kg (±0.07 x 10^9) for CST and SRF were obtained, respectively. In this way, the 2-phase anaerobic digestion presented a better dewatering trend compared to the single phase experiment. In fact, this improvement is largely due to the TS content on this digestate (Table 4).

Table 4: CST and SRF tests performed to assess the sludge dewaterability

Sample	Polymer dosage	SRF (mg/kgTS)	CST (s)
Raw sludge	0	1.64×10^8	139
Single phase digestate	0	Not filtered	Not filtered
	5.9	3.43×10^8	1737
	6.75	2.76×10^8	1095
	8.8	2.25×10^8	928
2-phase digestate	0	1.51×10^9	1602
	5.0	1.11×10^9	927
	8.7	9.94×10^8	984

4 Conclusions

The 2-phase anaerobic system showed clear increases in terms of organic matter removal and biogas production compared to the single stage system. The global SGP increased 32%, moving from 0.21 to 0.31 m^3/kgTVS.d for the single and 2-phase systems, respectively. A negative effect on WAS filterability was observed, where the single stage digestate presented the poorest dewatering trend. The heat produced from biogas in a CHP unit satisfied all the digester heat requirements besides producing an important surplus as renewable electrical power (2.5 to 3.0 MWh/d). The hydrolytic reactor payback time was estimated in 3 years.

5 Acknowledgments

Authors would like to thank the DAAD Program "Higher Education Excellence in Development Cooperation (Exceed)" and Exceed Swindon that made the participation at the event possible. Besides, authors thank CAPES Foundation for the International Mobility Program 'Ciência sem Fronteiras', Federal University of Santa Catarina/BR e University of Verona/IT. Finally, authors thank the water utility Alto Trevigiano Servizi S.r.l. (ATS) and the Municipality of Treviso for hosting the pilot hall within the area of the municipal treatment plant.

6 References

[1]   I.S. Turonskiy, P.K. Mathai, Wastewater sludge processing, Wiley Interscience, New Jersey, 2006.

[2]   Y. Chen, J.J. Cheng, K.S. Creamer, Inhibition of anaerobic digestion process: a review, Bioresour. Technol., 2008, 99, 4044–4064.

[3]   N. Duan, B. Dong, B. Wu, X. Dai, High-solid anaerobic digestion of sewage sludge under mesophilic conditions: feasibility study, Bioresour. Technol., 2012, 104, 150 – 156.

[4]   H.Q. Ge, P.D. Jensen, D.J. Batstone, Temperature phased anaerobic digestion increases apparent hydrolysis rate for waste activated sludge, Water Res., 2011, 45 (4),1597 – 1606.

[5]   A. Schievano, A. Tenca, S. Lonati, E. Manzini, F. Adani, Can two-stage instead of one-stage anaerobic digestion really increase energy recovery from biomass?, Appl. Energy, 2014, 124, 335–342.

[6]   E.R. Viéitez, S. Ghosh, Biogasification of solid wastes by two-phase anaerobic fermentation, Biomass Bioenergy, 1999, 16, 299–309.

[7]   G. Luo, L. Xie, Q. Zhou, I. Angelidaki, Enhancement of bioenergy production from organic wastes by two-stage anaerobic hydrogen and methane production process, Bioresour. Technol., 2011, 102(18), 8700–8706.

[8]   D. Bolzonella, P. Pavan, M. Zanette, F. Cecchi, Two-phase anaerobic digestion of waste activated sludge: effect of an extreme thermophilic prefermentation. Ind. Eng. Chem. Res., 2007, 46, 6650–6655.

[9]   I. Ferrer, F. Vazquez, X. Font, Comparison of the mesophilic and thermophilic anaerobic sludge digestion from an energy perspective. J. Residuals Sci. Technol., 2011, 8(2), 81-87.

[10]   D.F. Lawler, Y.J. Chung, S-J. Hwang, B.A. Hull, Anaerobic digestion: effects on particle size and dewaterability, J. WPCF, 1986, 58(2), 1107–1117.

[11]   F. Cecchi, P. Pavan, A. Musacco, J. Mata-Alvarez, G. Vallini, Digesting the organic fraction of municipal solid waste: moving from mesophilic (37 °C) to thermophilic (55 °C) conditions. Waste Manage. Res., 1993, 11, 403–414.

[12]   APHA, AWWA, WEF. Standard Methods for the Examination of Water and Wastewater. 21st ed. Washington, DC, 2005.

[13]   IRSA-CNR, Metodi Analitici per i Fanghi, Quaderno 64, Istituto di Ricerca Sulle Acque – Consiglio Nazionale delle Ricerche, Roma, 2006.

[14]   D. Bolzonella, C. Cavinato, F. Fatone, P. Pavan, F. Cecchi, High rate mesophilic, thermophilic, and temperature phased anaerobic digestion of waste activated sludge: a pilot scale study, Waste Manage., 2012, 32(6), 1196–1201.

[15]   M. A. de la Rubia, L. I. Romero, D. Sales, M. Perez, Temperature conversion (mesophilic to thermophilic) of municipal sludge digestion, Environ. Energy Eng., 2005, 51(9), 2581–2586.

[16]   X. Liao, H. Li, Biogas production from low-organic-content sludge using a high-solids anaerobic digester with improved agitation, Appl. Energy, 2015, 148, 252–259.

[17]   D. Bolzonella, P. Pavan, P. Battistoni, F. Cecchi, Mesophilic anaerobic digestion of waste activated sludge: influence of the solid retention time in the wastewater treatment process, Process Biochem., 2005, 40, 1453–1460.

[18]   H. N. Gavala, U. Yenal, I. V. Skiadas, P. Westermann, B. K. Ahring, Mesophilic and thermophilic anaerobic digestion of primary and secondary sludge. Effect of pre-treatment at elevated temperature, Water Res., 2003, 37, 4561–4572.

[19]   I. A. Nges, J. Liu, Effects of solid retention time on anaerobic digestion of dewatered-sewage sludge in mesophilic and thermophilic conditions, Renew. Energy, 2010, 35, 2200–2206.

[20]   C. M. Braguglia, N. Carozza, M. C. Gagliano, A. Gallipoli, A. Gianico, S. Rossetti, J. Suschka, M. C. Tomei, G. Mininni, Advanced anaerobic processes to enhance waste activated sludge stabilization. Water Sci.Technol., 2014, 69(8), 1728–1734.

INTERVAL APPROACHES FOR THE OPTIMIZATION AND CONTROL OF WASTEWATER TREATMENT PROCESSES BY ANAEROBIC DIGESTION

[1]Víctor Alcaraz-González, [1]Victor González Álvarez

[1] Departamento de Ingeniería Química, Universidad de Guadalajara, Blvd. Marcelino García Barragán 1451, C.P. 44430, Guadalajara Jalisco, México, victor.alcaraz@cucei.udg.mx

Keywords: Anaerobic Digestion, Interval Observers, Process Control, Robustness, Wastewater Treatment.

Abstract

Several approaches for the modeling, optimization and process control of anaerobic digestion for the treatment of municipal and agricultural wastewater are reviewed in this paper. In particular, anaerobic digestion is presented as study case for these approaches. Anaerobic digestion is known for showing instability under certain conditions. In this paper, at least one suitable approach is presented for dealing with some of these drawbacks in particular and for optimizing the whole process.

1 Introduction

Anaerobic Digestion (AD) processes refer to diverse arrays of biological wastewater treatment systems from which dissolved oxygen and nitrate are excluded. They are operated to convert biodegradable organic matter to biogas, which may include H_2, CH_4 and CO_2. Two areas, which are receiving a high level of international research, are: activated sludge management [1], and full recovery of resources, including water, nutrients, and energy. AD is a core component of this new concept, as the key biochemical process for mobilization of energy and nutrients. In particular, AD is used for treating wastewater with high organic loads, which causes several environmental damages when it is no adequately treated [2, 3]. But, on the other hand, AD exhibits some inconvenient features, like inhibition for substrate in the methanogenic phase, operational instability for alkalinity decompensating, among other troubles [4]. Based on the most recent literature and reviews (*e.g.*, [4] and the references cited therein), this paper aims to present some solutions to the most common drawbacks in AD. It is important to notice that there exist a number of different approaches for each situation or trouble, but this paper is far away to be a review. Instead of that, this paper presents a useful overview of 15 years authors' work and experience on the subject. Approaches introduced here can be implemented easily and at relatively low costs [4, 5]. Moreover, it is also shown that some of these approaches are robust against several problems like a full lack of knowledge on the kinetic parameters, uncertainty on the process inputs, and lack of sensors. Even in the presence of these troubles, the different approaches reviewed in this paper are able to reduce the residence time, to avoid some phenomena like washout of the biomass, and to avoid the inhibition for substrate among others. Because of space reasons, formal hypotheses needed for design and approaches building, as well as mathematical proofs are omitted. However, interested readers may find this information in the cited literature.

2 Materials and Methods

2.1 Anaerobic digestion model

AD includes a series of several physicochemical and biological reactions for degrading macromolecular organic matter into CH_4 and CO_2 as main final products [6, 7]. Different stages are involved and current models may be extremely complex (see *e.g.* Model ADM1 [8]), or simpler and more practical for process control purposes. In this paper, the AM2 model, developed and successfully validated in [5], is taken into consideration. It describes an AD process carried out in a continuous packed bed upper-flow digester. This model distinguishes the two most important stages of the processes: The first one is acidogenesis, where the main substrate, *i.e.*, organic matter, expressed as Chemical Oxygen Demand (COD), is degraded into Volatile Fatty Acids (VFA), and then the second one, called methanogenesis, transforms VFA into CH_4 and CO_2. It includes also the physicochemical phenomena in the reactor that play and important role in the bioreactor's operational stability. This model has been successfully applied for estimating unmeasured state variables, for process control and optimization in several actual AD processes [9-14]. AM2 model is depicted by equations (1a), where $x_1 \equiv X_1$, $x_2 \equiv X_2$ are the concentrations of acidogenic and methanogenic bacteria, and

$x_3 \equiv Z$, $x_4 \equiv C_{TI}$, $x_5 \equiv S_1$, $x_6 \equiv S_2$ are the concentrations of strong ions, total inorganic carbon, COD and VFA respectively. P_{CO_2} is the partial pressure of CO_2. The constant parameter α, $0 \le \alpha \le 1$ denotes the biomass fraction, which is retained by the reactor bed, *i.e.*, $\alpha = 0$ for the ideal fixed-bed reactor and $\alpha = 1$ for the ideal continuous stirred tank reactor. In all cases, the upper index "*in*" indicates "influent concentrations". The definition of all parameters and their values can be found in [15] and [16]. D is the dilution rate, whiles the specific growth rate μ_1 and μ_2 are given by Monod and Haldane functions, respectively, as depicted by equations (1b).

$$
\begin{bmatrix} \dot{x}_1 \\ \dot{x}_2 \\ \dot{x}_3 \\ \dot{x}_4 \\ \dot{x}_5 \\ \dot{x}_6 \end{bmatrix} =
\begin{bmatrix} 1 & 0 \\ 0 & 1 \\ 0 & 0 \\ k_4 & k_5 \\ -k_1 & 0 \\ k_2 & -k_3 \end{bmatrix}
\begin{bmatrix} \mu_1 x_1 \\ \mu_2 x_2 \end{bmatrix} +
\begin{bmatrix}
-\alpha D & 0 & 0 & 0 & 0 & 0 \\
0 & -\alpha D & 0 & 0 & 0 & 0 \\
0 & 0 & -D & 0 & 0 & 0 \\
0 & 0 & k_7 & -(D+k_7) & 0 & -k_7 \\
0 & 0 & 0 & 0 & -D & 0 \\
0 & 0 & 0 & 0 & 0 & -D
\end{bmatrix}
\begin{bmatrix} x_1 \\ x_2 \\ x_3 \\ x_4 \\ x_5 \\ x_6 \end{bmatrix} +
\begin{bmatrix} Dx_1^{in} \\ Dx_2^{in} \\ Dx_3^{in} \\ Dx_4^{in} + k_7 k_8 P_{CO_2} \\ Dx_5^{in} \\ Dx_6^{in} \end{bmatrix}
\qquad (1a)
$$

$$
\mu_1 = \frac{\mu_{\max 1} S_1}{K_{S_1} + S_1} \quad \text{and} \quad \mu_2 = \frac{\mu_0 S_2}{K_{S_2} + S_2 + \left(S_2 / K_{I_2} \right)^2} \qquad (1b)
$$

The former time-varying nonlinear system can be rewritten in the following compact form:

$$\dot{x}(t) = Cf(x(t),t) + A(D(t),t)x(t) + b(t) \tag{2}$$

where $x(t) \in \Re^n_+$ represents the state vector, $f(x(t),t) \in \Re^r_+$ is a positive vector that lumps the process kinetic terms, matrix $C \in \Re^{nxr}$ is a coefficients matrix (*i.e.*, yield coefficients), $A(t) \in \Re^{nxn}$ denotes the state matrix of the linear part of the system and $D(t) \in \Re_+$ is the dilution rate. Finally, $b(t) \in \Re^n$ is a vector that gathers the process inputs (*i.e.* the mass feeding rate vector) and other functions possibly time-varying (*e.g.* the gaseous outflow rate vector, if any). Now let us consider that only m states are measured on-line and, l of them are the states to be regulated $(l \leq m)$. Then, the state space can be split in such a way that (3) can be rewritten in measurable and non-measurable state variables as follows:

$$\dot{x}_1(t) = C_1 f(x(t),t) + A_{11}(t)x_1(t) + A_{12}(t)x_2(t) + b_1(t) \tag{3}$$

$$\dot{x}_2(t) = C_2 f(x(t),t) + A_{21}(t)x_1(t) + A_{22}(t)x_2(t) + b_2(t) \tag{4}$$

where the m measured states have been included in $x_2(t) \in \Re^m_+$ and the variables to be estimated are represented by $x_1(t) \in \Re^s$ $(s = n - m)$. Matrices $A_{11}(t) \in \Re^{sxs}$, $A_{12}(t) \in \Re^{sxm}$, $A_{21}(t) \in \Re^{mxs}$, $A_{22}(t) \in \Re^{mxm}$, $C_1(t) \in \Re^{sxr}$, $C_2(t) \in \Re^{mxr}$, $b_1(t) \in \Re^s$ and $b_2(t) \in \Re^m$ are the corresponding partitions of $A(t), C$ and $b(t)$, respectively.

2.2 Asymptotic and Interval Observers
There exist three important challenges in the control of WWT processes: a) is the lack of suitable online sensors able to provide fast and reliable measures of the key process variables. Actually, a current and very active research area in process control in general is the design of state observers for estimating unmeasured variables. b) the lack of a reliable knowledge or even a full ignorance on process kinetic parameters. These uncertainties are the consequence of the high diversity of microbial species in anaerobic sludges. In general, it is only possible to have an idea of the kinetic functions form (*e.g.*, Monod and/or Haldane type). The most classical approaches like the well-known Kalman Filter or the Luemberger Observer need this information and are not suitable enough to face this lack of knowledge. Thus, in this paper the asymptotic observer is considered as main tool for estimating state variables because it has the important feature to be fully independent of the nonlinear terms (*i.e.*, kinetics) [17, 15]. When $b(t)$ is known and there is no restriction on the initial conditions, the following system (5) is a stable asymptotic observer for the nonlinear time varying lumped model (4-5), *i.e.*, $\hat{x}_1(t)$ converges asymptotically towards $\mathbf{x_1(t)}$ for any initial conditions [15].

$$\begin{cases} \dot{\hat{w}}(t) = W(t)\hat{w}(t) + X(t)x_2(t) + Nb(t) \\ \hat{w}(0) = N\hat{x}(0) \\ \hat{x}_1(t) = N_1^{-1}\left(\hat{w}(t) - N_2 x_2(t)\right) \end{cases} \qquad (5)$$

$$\text{with} \quad W(t) = \left(N_1 A_{11}(t) + N_2 A_{21}(t)\right)N_1^{-1}, \quad X(t) = N_1 A_{12}(t) + N_2 A_{22}(t) - W(t)N_2$$

where $N_1 \in \mathfrak{R}^{s \times s}$ can be arbitrarily chosen with the only restriction to be non-singular. $N_2 \in \mathfrak{R}^{s \times r}$ is calculated as $N_2 = -N_1 C_1 C_2^{\perp}$ where $C_2^{\perp}$ is a generalized pseudo-inverse of C_2 such that $C_2^{\perp} = C_2 C_2^{\perp} C_2$ and finally $N = [N_1 \vdots N_2]$. Without loss of generality, N_1 will be chosen as the identity matrix $I \in \mathfrak{R}^{s \times s}$.

The third big challenge in control of WWT processes is the lack of knowledge on the influent concentrations. They can vary for seasonal reasons or even for human activities during the day. Moreover, very often only offline measurements are taken one time per day in best case. Nevertheless, if $b(t)$ is unknown but at least lower and upper bounds are known, the structure of the asymptotic observer (5) can be used to build an interval observer, *i.e.*, an observer that does not estimate the exact value of the unmeasured variables but a guaranteed finite interval, in which they evolve [18]. Thus, using the kinetic independent structure of (5), the following set-valued observer guarantees that $x_1^-(t) \le x_1(t) \le x_1^+(t), \forall t \ge 0$ given $x_1^-(0) \le x_1(0) \le x_1^+(0)$ [15, 19]:

For the upper bound:

$$\begin{cases} \dot{w}^+(t) = W(t)w^+(t) + X(t)x_2(t) + Mv^+(t) \\ w(0)^+ = N x(0)^+ \\ \hat{x}_1^+(t) = N_1^{-1}\left(w^+(t) - N_2 x_2(t)\right) \end{cases}$$

For the lower bound:

$$\begin{cases} \dot{w}^-(t) = W(t)w^-(t) + X(t)x_2(t) + Mv^-(t) \\ w(0)^- = N x(0)^- \\ \hat{x}_1^-(t) = N_1^{-1}\left(w^-(t) - N_2 x_2(t)\right) \end{cases} \qquad (6)$$

with $M = [N_1 \vdots N_2 \vdots \widetilde{N}_2]$, $\widetilde{N}_2 = [|N_{2,ij}|]$,

$$v^+(t) = \left[b_1^+(t) \quad \tfrac{1}{2}\left(b_2^+(t) + b_2^-(t)\right) \quad \tfrac{1}{2}\left(b_2^+(t) - b_2^-(t)\right) \right]^T,$$

$$v^-(t) = \left[b_1^-(t) \quad \tfrac{1}{2}\left(b_2^+(t) + b_2^-(t)\right) \quad -\tfrac{1}{2}\left(b_2^+(t) - b_2^-(t)\right) \right]^T$$

2.3 Control objectives: The operational stability

It is well known that the main cause of instability in anaerobic digestion is the accumulation of VFA that inhibits methanogenic bacteria (substrate inhibition). However, it is also well known that it is not enough to regulate only COD and VFA in order to avoid this trouble because the presence of CO_2 together with VFA also plays an important role on the physicochemical equilibrium. Thus, instead of only regulating VFA and or pH, the criterion most usually used is to regulate both, Total Alkalinity (TA) and Intermediate Alkalinity (IA). However, strictly speaking, they do not appear as state variables in the AM2 model (1). Nevertheless, TA and IA may be related to the true state variables S_2 and Z in the following form [16]:

$$TA \equiv f_{Tc}Z(t) + (f_{Ta} - f_{Tc})F_a S_2(t) \tag{7}$$

$$\frac{IA}{TA} \equiv \frac{f_{Ic}Z(t) + (f_{Ia} - f_{Ic})F_a S_2(t)}{f_{Tc}Z(t) + (f_{Ta} - f_{Tc})F_a S_2(t)} \tag{8}$$

with

$$f_{Tc} = \left(1 - \frac{10^{-pH(t)} + K_c}{10^{-4.3} + K_c}\right) \qquad F_a = \left(\frac{K_a}{10^{-pH(t)} + K_a}\right) \qquad f_{Pc} = \left(1 - \frac{10^{-pH(t)} + K_c}{10^{-5.75} + K_c}\right)$$

$$f_{Ta} = \left(1 - \frac{10^{-pH(t)} + K_a}{10^{-4.3} + K_a}\right) \qquad f_{Ic} = f_{Tc} - f_{Pc} \qquad f_{Pa} = \left(1 - \frac{10^{-pH(t)} + K_a}{10^{-5.75} + K_a}\right)$$

$$f_{Ia} = f_{Ta} - f_{Pa}$$

Thus, a number of practical operational criteria have been established for TA and the ratio IA/TA. For instance in [20], these criteria have been established as follows:

$$TA \geq TA_{\min} \equiv 60\,mEq/l \tag{9}$$

$$IA/TA \leq (IA/TA)_{\max} \equiv 0.3 \tag{10}$$

As a result, the control objectives may be stated as

$$S_2^r(t) \leq \frac{TA_{\min}\left(f_{Ic} - (IA/TA)_{\max} f_{Tc}\right)}{F_a\left(f_{Tc}\left((IA/TA)_{\max}(f_{Ta} - f_{Tc}) - (f_{Ia} - f_{Ic})\right) + (f_{Ta} - f_{Tc})\left(f_{Tc} - (IA/TA)_{\max} f_{Tc}\right)\right)} \tag{11}$$

$$Z^r(t) \geq \frac{TA_{\min} - F_a\left(f_{Ta} - f_{Tc}\right)S_2^r(t)}{f_{Tc}} \tag{12}$$

2.4 Control law

Finally in this subsection, a control law that exponentially stabilizes the output $y_i(t)$ around its set-point y_i^r is given by the following equation

$$D_i^*(t) = \frac{R_i^*(t) - \lambda_i^* \upsilon_i^*(t)}{\left(y_i^{in*}(t) - y_i^*(t)\right)} \tag{13}$$

where $y_i^*(t)$ may be COD, VFA or Z concentrations, which is measured or may be estimated by an interval observer like (6) or another classical observer like a Kalman or Luemberger observer. In the present case, the Luemberger one was used [16]. For each case, $R_i^*(t)$ represents the best available kinetic model taken into account the uncertainties of the kinetic parameters, while $y_i^{in*}(t)$ represents the best information on lower and upper values according to the case at each variable concerning process input concentrations. $v_i^*(t)$ represents a suitable Lyapunow function which is in concordance with each control objective and $\lambda_i^*(t)$ represents a gain that can be adjusted by the user [15, 16].

3 Results and Discussion

In this section, two main results are presented as example of the use of the control law (13) that at its time use all the other tools introduced in the former section. These study cases are the Single-Input-Single-Output (SISO) regulation for the COD [15] and the Multiple-Input-Multiple-Output (MIMO) regulation for alkalinity criteria [16].

3.1 SISO COD regulation

The experimental application of this regulation law was carried out in a 1 m^3 upflow anaerobic fixed bed reactor for the treatment of industrial wine distillery vinasses obtained from local distilleries in the Narbonne, France area. The objective was to lead $S_1(t)$ around a set-point but inside a reference interval tube.

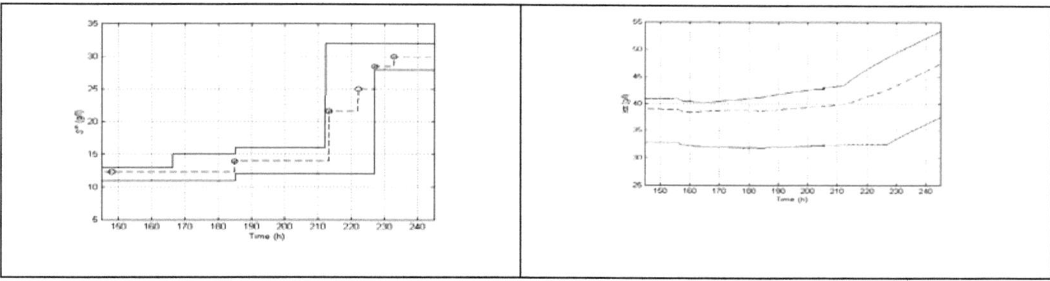

Figure 1:   Uncertainties in the influent COD concentration and interval estimate of the acidogenic concentration [15]

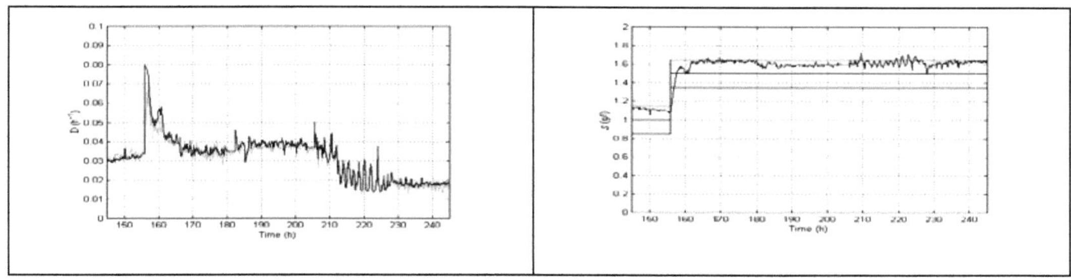

Figure 2:   Experimental results in the regulation of COD in an anaerobic digestion process using interval observers [15]

The regulation law was tested in the face of drastic perturbations induced in the organic load (see Figure 1). In this figure, the considered interval on $S_1^{in}(t)$ is represented by the continuous line (—). Off-line measurements (-o-) were taken sporadically only to validate the hypothesis considered on the input concentration interval. On the other hand, the product $k_1 X_1(t)$ is estimated by the interval observer (7) (continuous line —). The prediction of the model about $k_1 X_1(t)$ (dashed line --) is displayed here only to provide some kind of validation to the interval observer. Figure 2 shows the results obtained for the regulation of $S_1(t)$ using D as control input, which was calculated by the control law (13). The black line represents the signal given by (13) and the gray line represents the true signal given by the actuator.

The robustness of the regulation law in the face of load perturbations on $S_1^{in}(t)$ was performed as long as the set point was kept constant. Two perturbations operated in this time period. The first was a forced change of the feeding barrel at time t=166 h. As the corresponding new nominal value for $S_1^{in}(t)$ was only determined at t = 185 h, the uncertainty augmented from 11< $S_1^{in}(t)$< 13 to 11< $S_1^{in}(t)$<15 g/l during 166 < t < 185 h. Once the new nominal value was determined, the uncertainty was established as 12 < $S_1^{in}(t)$< 16 g/L. Up to this point, the reactor was fed with water-diluted vinasses (50:50 dilution). The second perturbation then consisted of allowing the reactor to be fed with pure vinasses at time t = 212 h. During the transition time period the uncertainty drastically increased (12< $S_1^{in}(t)$< 32 g/L) and induced an abrupt increased of the estimated interval for the product $k_1 X_1(t)$. Thus, the double enlargement in both, the uncertainty of $k_1 X_1(t)$ and the uncertainty on $S_1^{in}(t)$ forced the regulation law to compensate this major uncertainty with a bang-bang behavior, which, however, was correctly attenuated by the control law (14). This bang-bang behavior disappeared once, the uncertainty of $S_1^{in}(t)$ was reduced. Furthermore, in both perturbations, the regulation law performed very well satisfactorily keeping $S_1(t)$ inside the reference interval tube.

3.2 Alkalinity regulation

In order to show the versatility of this approach and how the operational stability can be easily manipulated by the user, a flexible operation was allowed by fixing the control objectives (9)-(10) as inequalities in the whole sense (*i.e.*, greater/smaller or equal than) and this was applied also for the regulation of Z (*i.e.*, $Z \geq Z^r$). Experimental runs were carried in the same pilot plant depicted in the former study case over a 12 d time period. Notice that only S_2 was regulated strictly around its set-point S_2^r. Figure 3 shows the constant references (dashed line) (that actually are a function of pH) as well as the measured variables S_2 and Z

(continuous line). The time evolution of the dilution rates D_i (the regulation laws), expressed as input flow rates, was calculated using the control law (13) using adequate Lyuapunov functions [16]. Experimental runs show how the control laws (13) is able to regulate the effluent concentrations S_2 and Z around the desired set-points S_2^r, and Z^r fulfilling in this form the operational stability criteria (9)-(10). Different set-points were used as shown in Table 1. The objective was to regulate the ratio IA/TA and TA as $(IA/TA)^r \leq 0.3$ and $TA^r \approx 90$ mmol/L, respectively. Notice that these control objectives were obtained in an indirect way by regulating the true states variables S_2 and Z.

Table 1: Set points used in the experimental phase

Period	1	2	3	4
TA (mmol/L)	90	90	85	75
IA/TA	0.3	0.28	0.27	0.3

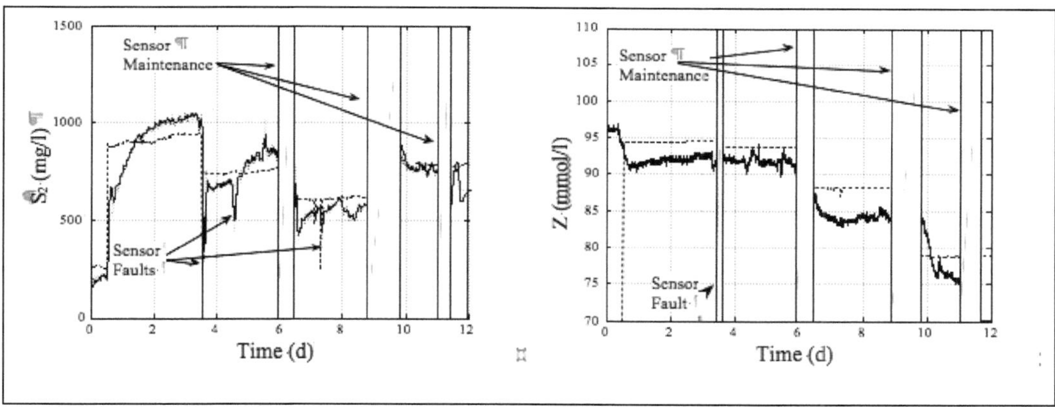

Figure 3: Regulation of S_2 and Z

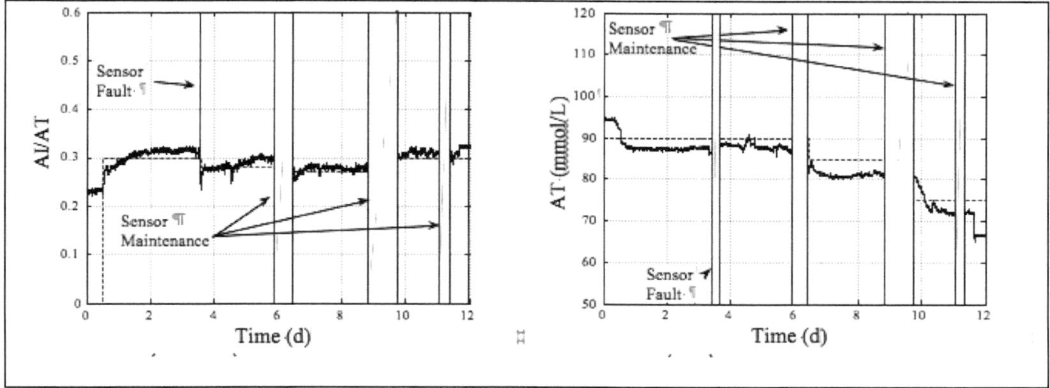

Figure 4: Regulation of IA/TA and TA

4 Conclusions

In this work, the automatic process control of an up-flow fixed bed continuous anaerobic digestion process has been carried out for the SISO and MIMO regulation of its main variables: the total organic load in the form of Chemical Oxygen Demand, and alkalinities, which are known as responsible for the process operational stability. The methodology used for archiving these objectives was based on interval observers and robust adaptive control approaches that take into account all the possible uncertainties in the process; *i.e.*, on the kinetics, on the process inputs, and even in the lack of measurements of variables involved on the control laws. These approaches were experimentally implemented and validate in an anaerobic digester for the treatment of red wine vinasses. Results have shown that these approaches are able not only to lead the regulated variables to its respective set points, and to show stability features, as theoretically expected, but also to a very high robustness against these uncertainties in actual conditions that demonstrate to be very useful for regulating this kind of processes.

5 Acknowledgements

This work was partially supported by Projects CONACyT CB-2008-01/101971, FORDECYT/ 116655. Authors also thanks DAAD and the ex)(ceed II / Swindon Project for financial support provided and for making this workshop possible.

6 References

[1]   D. J. Batstone, S. Tait, D. Starrenburg, Estimation of hydrolysis parameters in full-scale anaerobic digesters, Biotechnol. Bioeng. 2009, 102, 1513–1520.

[2]   D. Dochain, M. Perrier, Monitoring and Adaptive Control of Bioprocesses. In J.F.M. Van Impe, P.A. Vanrolleghem, D.M. Iserentant (Ed). Advanced Instrumentation, Data Interpretation, and Control of Biotechnological Processes. Kluwer Academic Publishers, Netherlands. 1998, p. 388.

[3]   J. Ph. Steyer, O. Bernard, D. Batstone, I. Angelidaki, Lessons learnt from 15 years of ICA in anaerobic digestion processes, Water. Sci. Technol. 2006, 53/(4-5), 25-33.

[4]   J. Jimenez, E. Latrille, J. Harmand, A. Robles, J. Ferrer, D. Gaida, C. Wolf, F. Mairet, O. Bernard, V. Alcaraz-Gonzalez, H.O. Mendez-Acosta, D. Zitomer, D. Totzke, H. Spanjers, F. Jacobi, A. Guwy, R. Dinsdale, G. Premier, S. Mazhegrane, G. Ruiz-Filippi, A Seco, T. Ribeiro, A. Pauss . J. Ph. Steyer, Instrumentation and control of anaerobic digestion processes: a review and some research challenges. Rev. Environ. Sci. Biotechnol. 2015, 14/4, 615-648.

[5]   O. Bernard, Z. Hadj-Sadok, D. Dochain, A. Genovesi, J.Ph. Steyer. Dynamical model development and parameter identification for anaerobic wastewater treatment process. Biotechnol. Bioeng. 2001, 75/4, 424–438.

[6]   M. Henze, P. Harremoës, J. LaCour, E. Arvin, Wastewater Treatment: Biological and Chemical Processes. Springer, Heidelberg 1995.

[7]   M. Henze, P. Harremoes. Anaerobic treatment of wastewater in fixed film reactors - a literature review. Water. Sci. Technol. 1983, 15/(8-9), 1-101.

[8]   [8] D. J. Batstone, J. Keller, I. Angelidaki, S. Kalyuzhnui, S. G. Pavlostathis, A. Rozzi, W. Sanders, H. Siegrist, V. Vavilin, Anaerobic Digestion Model No.1 (ADM1). IWA Publishing, London 2002.

[9]   R. Antonelli, J. Harmand, J. Ph., Steyer, A. Astolfi, Set-Point Regulation of an Anaerobic Digestion Process with Bounded Output Feedback. IEEE Trans. Control Syst. Technol. 2003, 11/4, 495–504.

[10]   V. Alcaraz-Gonzalez, J. Harmand, A. Rapaport, J. Ph. Steyer, V. Gonzalez-Alvarez, C. Pelayo-Ortiz, Robust interval-based regulation for anaerobic digestion processes. Water Sci. Technol. 2005, 52(1-2), 449-456.

[11]   F. Grognard, O. Bernard, Stability Analysis of a Wastewater Treatment Plant with Saturated Control. Water Sci. Technol. 2006, 53/1, 149–157.

[12]   H. O. Méndez-Acosta, B. Palacios-Ruiz, J. Ph. Steyer, V. Alcaraz-González, E. Latrille, V. González-Álvarez, Robust Control of Volatile Fatty Acids in Anaerobic Digestion Processes. Ind. Eng. Chem. Res. 2008, 47/20, 7715-7720.

[13]   H.O. Méndez-Acosta, B. Palacios-Ruiz, V. Alcaraz-González, V. González-Álvarez, J. P. Garcia-Sandoval, A Robust Control Scheme to improve the Stability of Anaerobic Digestion Processes. J. Process Control. 2010, 20, 375–383.

[14]   H. O. Méndez-Acosta, J. P. García-Sandoval, V. González-Álvarez, V. Alcaraz-González, J. A. Jáuregui-Jáuregui, Regulation of the organic pollution level in anaerobic digesters by using off-line COD measurements. Bioresour. Technol. 2011, 102, 7666-7672.

[15]   V. Alcaraz-González, V. Gonzalez-Álvarez, Robust Nonlinear Observers for Bioprocesses: Application to Wastewater Treatment. In H. O. Méndez-Acosta, R. Femat, V. González-Álvarez, V. (Ed). Selected Topics in Dynamics and Control of Chemical and Biological Processes, Springer, Berlin 2007, p. 125–172.

[16]   V. Alcaraz-González, F.A. Fregoso-Sánchez, H.O. Méndez-Acosta, V. González-Álvarez, Robust Regulation of Alkalinity in Highly Uncertain Continuous Anaerobic Digestion Processes. CLEAN – Soil, Air, Water 2003, 41/12, 1157- 1164.

[17]   D. Dochain, State and parameter estimation in chemical and biochemical processes: a tutorial. J. Process Control. 2003, 13/8, 801–818.

[18]   A. Rapaport, J. Harmand, Robust regulation of a class of partially observed nonlinear continuous bioreactors. J. Process Control. 2002, 12/2, 291–302.

[19]   V. Alcaraz, J. Harmand, A. Rapaport, J. P. Steyer, V. González, C. Pelayo, Software sensors for highly uncertain WWTPs: a new approach based on interval observers, *Wat. Res.* 2002, 36/10, 2515–2524.

[20]   O. Bernard, M. Polit, Z. Hadj-Sadok, M. Pengov, D. Dochain, M. Estaben, P. Labat, Advanced monitoring and control of anaerobic wastewater treatment plants: III software sensors and controllers for an anaerobic digester, in *Watermatex-2000*, 2000, Gent, Belgium, 3, 65–72.

SORPTION OF PHENOL/TYROSOL FROM NANOFILTRATE STREAM OF A PRETREATED OLIVE MILL WASTEWATER USING A MACRO-RETICULAR AROMATIC POLYMERIC RESIN

[1,2]Jacques Romain Njimou, [2]Marco Stoller, [2]Agnese Cicci, [2]Angelo Chianese, [1]Charles Péguy Nanseu-Njiki, [1]Emmanuel Ngameni, [2]Marco Bravi

[1] Laboratory of Analytical Chemistry, Faculty of Sciences, University of Yaoundé I, B.P. 812 Yaoundé, Cameroon
[2] Department of Chemical Materials Environmental Engineering, University of Rome "La Sapienza", Via Eudossiana 18 – 00184, Rome, Italy (njimoujacques@gmail.com)

Keywords: equilibrium, fixed-bed, macro-reticular aromatic, nanofiltrate, phenol/tyrosol

Abstract

Phenol and tyrosol are two phenolic compounds found in the effluents from olive mill waste waters (OMWW). Owing to its toxicity and its reactivity, phenol is slightly acidic, soluble in water. Likewise, tyrosol together with hydroxytyrosol belongs to the group of antioxidants contained in OMWW and deserves a close attention for the pharmaceutical, cosmetic and food industries. Polyphenol compounds, lipid and organic acid may be transformed into phytotoxic materials. This is the reason why the disposal of the OMWW into agricultural ground, usually adopted in the Mediterranean countries, cannot be adopted any more. In the current work, the concentrate of nanofiltration containing the low-molecular-weight compounds of the waste occurring from a membrane process was used for further treatment with R&H FPX66 aromatic resin. The operating parameters affecting the adsorption in batch mode process were studied and phenol adsorption was found to be larger than that of tyrosol. Equilibrium data was properly fitted by a pseudo-second-order rate fitting equation. The non-ionic FPX66 resin was implemented in a fixed bed column, for the recovery of phenols and their separation from compounds of interest.

1 Introduction

Olive mill waste waters (OMWW) represent an important environmental problem in Mediterranean areas where they are generated in huge quantities in short periods of time. The high concentration in phenolic compounds (4-15 g/L), lipid and organic acid make them phytotoxic materials. But these wastes also contain valuable resources such as a large proportion of organic matter and a wide range of nutrients that could be recycled [1]. The seasonal pollution load of olive-oil production in these regions is equivalent to the one of 22 million people, since in average of the total chemical oxygen demand (COD) of OMWW is about 80 g.L^{-1} and its quantity is about 0.8 ton per ton of olive. Moreover, this huge amount of polluted wastewater is produced only during 2-3 months per year (average duration of the olive oil campaign), making the storage difficult and entailing the need of an immediate treatment to eliminate its environmental hazard. The discharge of OMMW is not allowed through the municipal sewage system and/or natural effluents. At the same time, however, no efficient and low cost treatment exists for the treatment of such effluents to be performed in an industrial environment such as the mill factory. For this reason, the current disposal systems adopted, which in most cases infringe the European and national regulations, OMWW are mostly evaporated and disposed of on lands. The problems caused by this procedure are pollution of groundwater layers and of terrains. For this reason, the procedure is considered illegal according to the EU Directive 91/271/EEC on urban wastewater treatment. This wastewater contains many polyphenols, mostly hydroxytyrosol, which may have a market value [2–4]. For many reasons, it is not easy to extract directly polyphenols from raw wastewater. The high suspended solids in the raw stream will quickly block the adsorption column, making the recovery process impossible. Moreover, many other interfering and undesired pollutants would interfere with the recovery process. For this reason, the recovery of polyphenols was accomplished on a pretreated stream that is the fraction of nanofiltration produced after raw wastewater treatment by flocculation, photocatalysis and ultrafiltration. The concentrate of nanofiltration (NF) is rich enough of polyphenols to permit a suitable recovery from technical and economic point of view. Finally, both batch and fixed bed modes adsorption of phenol/tyrosol in aqueous solution and nano-filtrate stream from pretreated OMWW by macro-reticular aromatic polymeric resin will be presented.

2 Material and Methods

2.1 OMWW and pretreatment processes

OMWW in 30 L bottles were transported to the laboratory and kept at - 20 °C for later use. The pretreatment involved the following: OMWW sieving was operated using a 300 micron sieve. Before flocculation using aluminum sulphate, acidification was performed using a nitric acid, and its initial pH 5.2 was reduced to pH 3.0. Then the mixture was centrifuged and the supernatant was exposed to UV irradiation. The photocatalysis is an innovative and promising technique for the purification of OMWW. It is well known that titanium dioxide (TiO_2) is a catalyst widely used due to its high performance, low cost, high photo-activity, low toxicity, chemical stability, insolubility and the resistance to photo-corrosion. The clarified stream was processed by photocatalysis using home-made composite magnetic nanoparticles with size of 79 nm coated by titania, irradiated by a UV lamp VL-315 BLB 3x15W- 385 nm Tube Power 90W for 4 hours [5, 6], and the membrane filtration experiments were carried out immediately after sample collection.

2.2 Membrane processes

Although the results from membrane filtration have already been published [2], they are briefly presented in this section in order to facilitate presentation of current work. The clarified OMWW was ultra-filtered on GM membranes, type (TFM) tubular with the cut-off between 100 and 200 kDa, and the recovering of permeate.

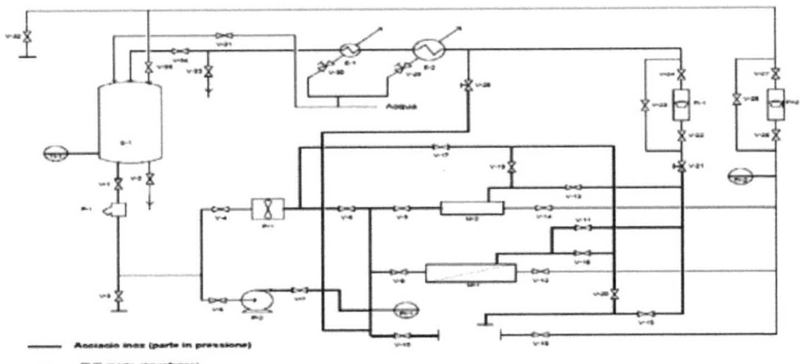

Figure 1: Schematic of pilot plant used «Ecosystem S. L Technologies»

The ultrafiltration method was followed by the nanofiltration on the Osmonics Model DK, spiral and tubular membranes type (TFM) with the cut-off between 150 and 400 Da, and at the maximum pressure of 16 bar.

The pilot plant used in Figure 1 was an Ecosystem S. L Technologies and constituted of:
S_{01}: Feed tank with the capacity: min 10 L, max 100 L where the water to be treated was introduced; F_{01}: Cartridge filters of 50 μm to remove power remaining solids; P_{01}: Volumetric pump; P_{02}: Centrifuge pump; M_{01}: Housing for medium size membrane modules, mod. 2540, area 2.51 m^2; M_{02}: Housing for small size membrane modules, mod. 1812, area 0.52 m^2; Fl_{01}: Flowmeter concentrate stream; Fl_{02}: Flowmeter permeate stream; E_{01}: concentrate heat exchanger (big); E_{02}: concentrate heat exchanger (small); V_{xx}: valves (x = 1, 2, 3... 35). The centrifugal booster pump P_1 and the volumetric pump P_2 drive the wastewater stream over the spiral-wound membranes used, supplied by Osmonics, fitted in the housing M_1, at an average flow rate of 600 L/h, the active area of each module membrane was equal to 2.51 m^2 [2, 4, 7, 8].

By acting on the regulation valves V_1 and V_2, it was possible to set the desired operating pressure P_{ExT} over the membrane to maintain a constant feed flow rate with an accuracy of 0.5 bar. Both permeate and concentrate streams were cooled down to the feedstock temperature, mixed together and recycled back to the feedstock, and the feedstock composition was maintained constant during each experimental batch run. The temperature was set at 20 °C ± 1 °C for all experiments [2– 4]. The pretreatment processes aimed at reducing the total suspended solids (**TSS**) and organic matter by measuring the **COD**. Total chemical oxygen demand (**COD**) was measured according to the 5220 D method, and total suspended solids (**TSS**) according to the 2540 D of the Standard Methods [2]. Phenols (Ph) were measured with the Folin–Ciocalteu method [2], using gallic acid as standard at 760 nm.

Table 1: Characteristics of membrane and the pretreatment of OMWW

Type	Characteristic of membrane modules				
	Id	m_w [L/hm^{2}bar]	m_w [L/hm^{2}bar]	D_p [nm]	P_{max} [bar]
UF	Osmonics model GM	16.3	4.8	2.0	16
NF	Osmonics model DK	7.9	5.2	0.5	32
RO	Osmonics model SC	2.7	2.6	< 0.1	65
Stream		Pretreatment of OMWW raw materials			
		COD		**TSS**		**pH**
		[g/L]	Δ%	[g/L]	Δ%	-
Raw OMWW		32.4	-	33.0	-	5.2
After flocculation		22.2	31.5	10.9	66.9	3.1
After centrifugation		19.2	13.5	8.4	22.9	-
After photocatalysis		16.5	14.1	5.2	38.1	-

The obtained results by authors [2– 4] are shown in Table 1 and reported as a percentage of the reduction (**Δ%**). Following this operation, the nanofiltrate fraction is rich in polyphenols and was used in the adsorption/desorption tank for phenolic separation on macro-reticular aromatic polymeric resin. FPX66 resin was provided by Rohm & Haas, and its physicochemical characteristics were white spherical beads, moisture holding capacity (60-80%), shipping weight of 680 g/L, surface area ≥ 700 m^2/g, specific gravity (1.015- 1.025), porosity ≥ 1.4 cc/g, and harmonic mean size (600 - 750 μm).

2.3 Adsorption and desorption on FPX66 resin in batch and continuous modes
In batch mode, 10 g of swollen resin with EtOH was suspended in a stirred tank reactor and beaker thermostatic batch conditions. 200 mL of phenol/tyrosol solutions of different concentrations (200 to 600 mg/L) were added for a certain time until thermodynamic equilibrium was reached. After this, the resin carrying the adsorbed polyphenols were recovered back (by gravity) and transferred into the desorption tank. In order to determine the exact concentration of phenol/tyrosol and to establish the evolution of the adsorption process, water samples from the supernatant were collected at different time intervals, till equilibrium was reached. To evaluate the residual concentrations of phenol and tyrosol after each adsorption/desorption experiment, the slurry was filtered through 0.45 μm polytetrafluoroethylene (PTFE) filters and the permeate was analysed by HPLC system composed of a Spectra-Physics liquid chromatography Model Agilent 1200 Series equipped with SUPELCOSIL LC-18 column (length 250 mm, diameter 4.6 mm, packaging size 5 mm). The mobile phases were acetonitrile (solvent A) and 0.5% (v/v) and acetic acid in water (solvent B) at a flow rate of 1mL/min. For each compound the adsorption capacity q, mg/g, the removal efficiencies (**E**, %) and the desorption ratio of the resin **D** (%) were calculated using Eqs. (1), (2) and (3), respectively [9, 10].

$$q = \frac{(C_0 - Ce)}{m} \times V \ \textbf{(1)}, \qquad E(\%) = \frac{C_0 - C_e}{C_0} \times 100 \ \textbf{(2)} \quad \text{and} \quad D(\%) = \frac{C_d V_d}{(C_0 - C_e)V} x\ 100(\%) \ \textbf{(3)}$$

Where C_o and C_e (mg/L) are the initial and equilibrium concentrations of analyte in the solutions, **m** (g) the quantity of the adsorbent, C_d (mgL^{-1}) is the concentration in the desorption solution; **V** and V_d (L) are the volume of the initial adsorption and desorption solutions, respectively.

Continuous fixed-bed adsorption runs were performed in glass columns reactor with an inner diameter of 4 cm and a length of 3.8 cm. The FPX66 resin was packed in the column with two layers of glass wool at the top and bottom as shown in Figure 2. The phenolic solution was charged from the top of the column in a down flow method at a fixed feed flow rate maintained throughout the experiments by using a peristaltic pump.

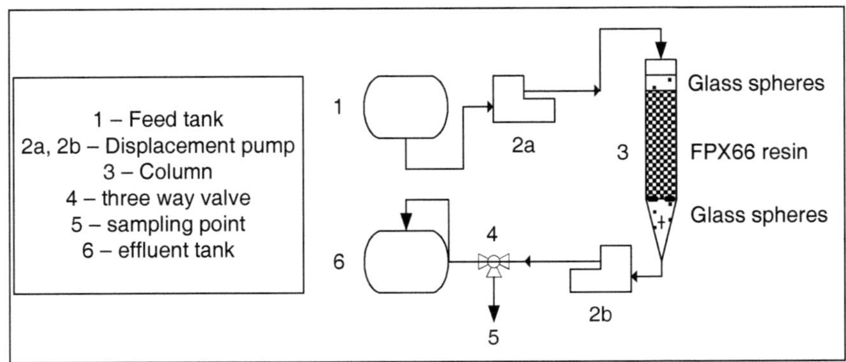

Figure 2: Schematic diagram of the experimental setup

With respect to the breakthrough separation of phenol, tyrosol and HO-tyrosol in binary and in OMWW NF concentrate, a tyrosol solution with 100% purity was represented by a time (h) in which the outlet contains only tyrosol since its breakpoint (B.T = Breakthrough time) is given as:

$$\Delta V100 \ (mL) \ = \ Q[time_{(100\%)} - time_{(B.T)}] \qquad \textbf{(4)}$$

A tyrosol solution with 90% purity was represented by a time (h), in which the outlet contains only tyrosol at 90% and phenol at 10% since its breakpoint (B.T = Breakthrough time) is given as:

$$\Delta V90 \ (mL) = \ Q[time_{(90\%)} - time_{(B.T)}] \qquad \textbf{(5)}$$

3 Results and Discussion

3.1 Sorption in batch mode

The effect of contact time on the adsorption/desorption of phenol/tyrosol onto FPX66 resin, respectively, was plotted at C_o = 500 mgL^{-1} against time as shown in Figure 3. The results indicated that the uptake of both solutes was rapid during the initial period of 45 min, and thereafter became slower near the equilibrium. This is apparent to the fact that a large number of vacant surface sites are available for the adsorption during the initial stage, whereas close to equilibrium the residual vacant sites are difficult to be occupied due to repulsive forces between the solute molecules on the solid and bulk phases [10, 11]. No significant change in phenol/tyrosol removal was observed after a contact time of 1 hour. It is observed that, the amount of phenol adsorbed were higher than those of tyrosol adsorbed for the FPX66 resin. The removal efficiencies (E, %) for the resin FPX66 was reported 57% for the tyrosol and 83% for the phenol at the same time. The adsorption of aromatic compounds as tyrosol/phenol on a non-functionalized support as FPX66 is a principle of combination of two forces: Van-der Waals forces and a thermodynamic gradient determined by the hydrophobicity, driving them out of the aqueous solution. Macro-reticular aromatic as well as macro-porous polystyrene cross-linked polymeric adsorbents usually have a high hydrophobic surface and a high affinity for adsorbate with phenyl groups. In fact, the fundamental interactions between the surface and aromatics compounds are the dispersion effect between the aromatic ring, the π-electrons [12], the electrostatic attraction and repulsion of present ions [13, 14].

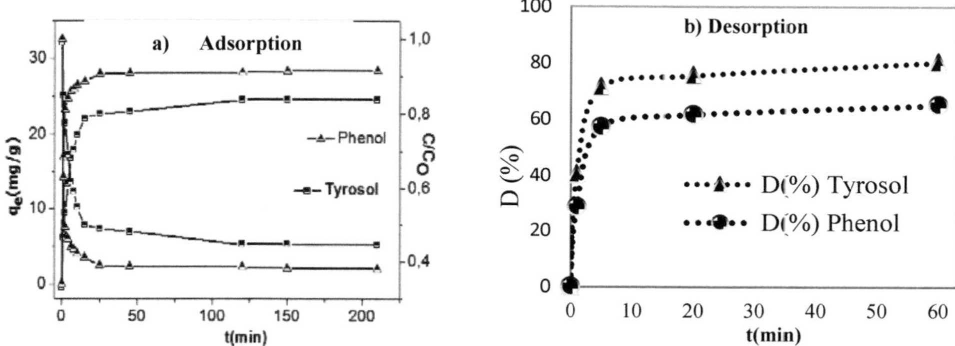

Figure 3: Concentration evolution (**C/C₀**, right side axis) and amount adsorbed (**q**, left side axis) (a); Evolution desorption percentage in EtOH milieu (b) of phenol/ tyrosol as function of time onto FPX66 resin

The reusability of an adsorbent is an important parameter to evaluate its economic feasibility in an adsorption process. The possibilities to reuse the FPX66 resin several times was undertaken by examining the phenol/tyrosol desorption in EtOH solution and the desorption percentages from FPX66 resin. The results are given in Figure 3-b, where the variation of desorption percentage is plotted against time. Desorption percentage obtained with 50% of EtOH in aqueous solution was very fast for FPX66 resin and the equilibrium was attained after 30 min. The results clearly show that tyrosol is adsorbed slightly and desorbed faster. FPX66 resin could be used many times without losing its significant adsorption capacities for the removal of both solutes. Close to 85% of phenol and 94% of tyrosol were desorbed on FPX66 by using the solution of 50 % EtOH, respectively. So, it is clear that desorption on FPX66 resin results from the cation exchange between the protons in solution and the adsorbed solute [15]. The non-adsorbed fraction of analyte may be bound to the resin through a different mechanism from cation exchange.

Phenol/tyrosol adsorption data on FPX66 resin were analysed by using a pseudo-first-order and pseudo-second-order kinetic model equations. Equilibrium data fitted good correlation with the coefficient ($R^2 \gg 0.99$) by a pseudo-second-order model showing that the adsorption of phenol/tyrosol on FPX66 resin followed the second order kinetic model. The calculated values of the solid solute concentration [($q_{e\ cal}$, mg/g): **Ph** (28.48) and **Ty** (25.38) on FPX66; Ph (29.09) were very close to those experimentally obtained [($q_{e\ exp}$, mg/g): **Ph** (28.88) and **Ty** (25.00) in single component system and [($q_{e\ cal}$, mg/g): **Ph** (38.31) and **Ty** (26.00), and [($q_{e\ exp}$, mg/g): **Ph** (38.19) and **Ty** (26.34) in binary components system, respectively.

3.2 Sorption in continuous mode

Based on the experimental data obtained in bicomponent phenol/tyrosol adsorption by FPX66 resin in column approach and reminding that the inlet concentration of tyrosol and phenol were equal, it can be concluded that tyrosol adsorption is weaker than phenol adsorption since the breakpoint of tyrosol occurs earlier than that of phenol. This sequence corresponds to the sequence of increasing affinity of this solute toward the macro-reticular aromatic (FPX66) polymeric resin. In fact, during the early stage of the column operation, the FPX66 resin bed possesses abundant adsorption sites for both incoming phenolic species.

Treating wastewaters containing two solutes phenol/tyrosol, it is highly desirable to separate and recover one of the components. Water recovery is also an important consideration because of the limited and expensive water resource. Adsorption by using standard solutions of phenol/tyrosol showed that the tyrosol or phenol concentrations in the first bed volume of effluent are almost zero for both solutes. This effluent can be discharged directly or recycled to be processed as supplement water. From the tyrosol breakpoint BV to solution of tyrosol with around 90% purity in BV, the effluent contains mainly tyrosol and can be sent to a unit for tyrosol recovery. For the effluent with about 90% of tyrosol BV, the FPX66 resin is saturated with the feedstock and needs regeneration to restore its adsorptive capacity.

In order to determine the effectiveness of adsorption in a more complex and realistic scenario, FPX66 resin in a fixed mode was exposed to an OMWW NF fraction containing all polyphenols from the membrane of nanofiltration plant selected for this work [3, 8, 16].The influent flow rate was kept at a constant flow rate of 2 mL/min whereas bed height, diameter of column and resin particle size were **Z** = 3.8 cm, **D** = 4 cm and 600 ≤ **d** ≤ 750 µm at ambient temperature, respectively.

Table 2 shows the resulting profiles, which gather the estimated breakthrough time and the adsorption capacities for the effluent containing two remarkable polyphenols selected in this work, such as tyrosol and OH-tyrosol, with purity up to 90%. This condition permitted a direct comparison with the data gathered in single and binary component adsorptions since the concentration swings were probably identical. A quick comparison between the breakthroughs revealed that a competitive adsorption is quite remarkable in all of the studies in FPX 66 resin.

Table 2: Experimental results by using a NF fraction of pretreated OMWW

Feed stream components		Tyrosol	OH-Tyrosol	Phenol
C$_{In}$ [mg/L]		171.12	123.01	122.55
Feed flow rate [mL/min]		2.0	2.0	2.0
BT	time [h]	6.5	8.5	22.5
	volume [ml]	780	1020	2700
Solution of Tyrosol at 100% purity in mass	time [h]	22.5	22.5	22.5
	volume [mL]	2700	2700	2700
	ΔV_{100}* [mL]	1920	1920	1920
	mass [mg]	174.29	87.02	0.00
	Con [mg/L]	90.78	51.80	0.00
Solution of Tyrosol at 90% purity in mass	time [h]	28.0	28.0	28.0
	volume [mL]	3360	3360	3360
	ΔV_{90}* [ml]	2580	2580	2580
	mass [mg]	229.35	137.15	9.27
	Con [mg/L]	88.90	53.16	3.40

A considerable reduction in breakthrough time was observed for tyrosol i.e. 12; 11.5 and 6.5 hours, respectively, in single, binary and multiple component systems. An increase of breakthrough time for the phenol during 17; 18 and 22.5 hours, respectively, proved that the selectivity for phenol observed was higher than of other polyphenols on FPX66 resin.

Desorption was carried out by using the solution of 50% ethanol under same conditions as adsorption column study, i.e. the influent flow rate was kept at a constant flow rate 2 mL/min, bed height, diameter and resin particle size were Z=3.8 cm, **D**= 4 cm and 600 ≤ **d** ≤ 750 µm at ambient temperature, respectively.

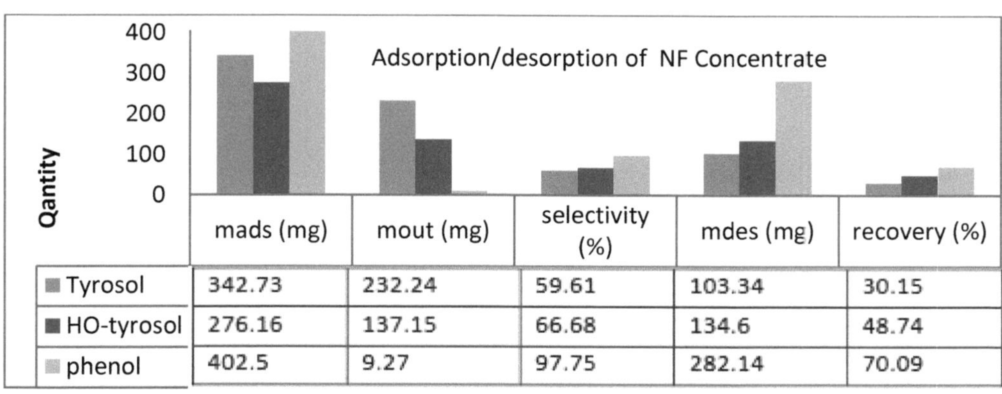

	mads (mg)	mout (mg)	selectivity (%)	mdes (mg)	recovery (%)
Tyrosol	342.73	232.24	59.61	103.34	30.15
HO-tyrosol	276.16	137.15	66.68	134.6	48.74
phenol	402.5	9.27	97.75	282.14	70.09

m$_{Ads}$: mass adsorbed; m$_{Des}$: mass desorbed; mass$_{sended}$ = m$_{Ads}$ + m$_{out}$

Figure 4: Chart of experimental data and mass balance obtained from the adsorption and desorption of the fraction of nanofiltration

Figure 4 presents the results obtained for both adsorption and desorption of OMWW NF and the mass balance was performed for three representative phenols, i.e., phenol, tyrosol and HO-tyrosol, and it also confirms the obtained results during adsorption of phenol/tyrosol in single and binary systems.

Apparently, phenol is more adsorbed on FPX66 resin with the nanofiltrate, which is a complex system compared to single and binary systems. The selectivity of phenol (97.75%) was greater than that of tyrosol (59.61%) and OH-tyrosol (66.68%). During the adsorption in the real system, the nanofiltrate could be explained by the hydrophobicity due to the preponderance of phenolic compounds. Desorption was faster compared to the adsorption. The ethanol solution of 50% allowed the recovery rates of 30.1, 48.7 and 70.1% of adsorbed tyrosol, HO-tyrosol and phenol, respectively.

4 Conclusions

The study showed that a macro-reticular aromatic can be used as an effective adsorbent for the removal of phenol/tyrosol from aqueous solutions. Quantitatively, the adsorption of phenol/tyrosol involves very fast kinetics of the pseudo-second order type. There was a greater adsorption of phenol compared with tyrosol, and desorption with ethanol permitted a bath recovery of a maximum of 84.6% of phenol and 94 % of tyrosol on FPX66 resin, respectively. Through the proposed procedure, the recovery of phenols in nanofiltrate and their separation from compounds of interest contained in the olive mill wastewater was possible through resin adsorption/desorption. During the intermediate stage of the column operation, adsorbed tyrosol molecules were replaced by the incoming phenol molecules. Tyrosol and phenol concentrations in the first bed volumes were 0 mg/L; this effluent could be discharged directly as water resource. From a tyrosol breakthrough of around 90% purity in bed volumes, effluent contains mainly tyrosol and could be sent to a unit for tyrosol recovery. For effluent above 90% of tyrosol in bed volumes, FPX66 resin was saturated with the feedstock and needed regeneration.

5 Acknowledgements

We wish to acknowledge the contribution of DAAD and Exceed Swindon that made the participation of this event possible, and the support received by the European Commission in the framework of Mundus ACP, which permitted this research at the Department of Chemical Engineering Materials Environment of the University of Rome "La Sapienza" during which the present scientific work was performed.

6 References

[1]   A. Roig, M. L. Cayuela, M. A. Sánchez-Monedero, An overview on olive mill wastes and their valorisation methods, Waste Manag. 2006, 26, 960–969.

[2]   A. Cicci, M. Stoller, M. Bravi, Microalgal biomass production by using ultra- and nano-filtration membrane fractions of olive mill wastewater, Water Res. 2013, 47, 4710–4718.

[3]   M. Stoller, On the effect of flocculation as pretreatment process and particle size distribution for membrane fouling reduction, Third Membr. Sci. Technol. Conf. Visegrad-Ctries. PERMEA, 2009, 240, 209–217.

[4]   M. Stoller and M. Bravi, Critical flux analyses on differently pretreated olive vegetation waste water streams: Some case studies, Desalinat. 2010, 250, 578–582.

[5]   Y. Ruzmanova, M. Ustundas, M. Stoller, A. Chianese, Chem. Eng. Transact. 2013, 2233–2238.

[6]   V. Vaiano, O. Sacco, M. Stoller, A. Chianese, P. Ciambelli, D. Sannino, Inter.J.Chem. Reactor Eng. 2014, 2013-0090.

[7]   M. Stoller, B. De Caprariis, A. Cicci, N. Verdone, M. Bravi, A. Chianese, About proper membrane process design affected by fouling by means of the analysis of measured threshold flux data, Sep. Purif. Technol. 2013, 114, 83–89.

[8]   M. Stoller, M. Bravi, A. Chianese, Spec. Issue Nanofiltration Membr. Fundam. Appl. 2013, 315, 142–148.

[9]   J. E. Mbosso T., A. Măicăneanu, C. Indolean, J. R. Njimou, C. Majdik, Cd^{2+} removal from aqueous solutions using an organo-inorganic immobilized adsorbent, Rev. Roum. Chim. 2012, 57, 321–325.

[10]   J. R Njimou., A. Măicăneanu, C. Indolean, C. P. Nanseu-Njiki, E. Ngameni, Removal of Cd (II) from synthetic wastewater by alginate - Ayous sawdust (Triplochitonscleroxylon) composite material, Environ. Technol. 2016, 1479–487X.

[11]   P. Barkakati, A. Begum, M. L. Das, P. G. Rao, Adsorptive separation of Ginsenoside from aqueous solution by polymeric resins, Chem. Eng. J. 2010. 161, 34–45.

[12]   M. Ahmaruzzaman and D. K. Sharma, Adsorption of phenols from wastewater, J. Colloid Interface Sci., 2005, 287, 14–24.

[13]   M. Gong, W. Zhu, Z. R. Xu, H. W. Zhang, H. P. Yang, Influence of sludge properties on the direct gasification of dewatered sewage sludge in supercritical water, Renew. Energy 2014, 66, 605–611.

[14]   W. M. Zhang, Q. J. Zhang, B. C. Pan, L. Lv, B. J. Pan, Z. W. Xu, Q. X. Zhang, X. S. Zhao, W. Du, Q. R. Zhang, J. Colloid Interface Sci. 2007, 306,216–221.

[15]   M. L. Soto, A. Moure, H. Domínguez, J. C. Parajó, A review, J. Food Eng. 2011, 105, 1–27.

[16]   J. M. Ochando-Pulido, M. Stoller, M. Bravi, A. Martinez-Ferez, A. Chianese, Sep. Purif. Technol. 2012, 101, 34–41.

[17]   K. Z. Setshedi, M. Bhaumik, M. S. Onyango, A. Maity, J. Ind. Eng. Chem.2014, 20, 2208–2216.

CATALYZING FENTON REACTION FOR DYEING COLOR DISPOSAL BY USING UNCOATED-POTTERY SCRAPS (UPS)

[1]Somphinith Muangthong, [2]Natkanin Supamathanon

[1]Rajamangala University of Technology Isan, Faculty of Engineering and Architecture, 744 Suranaray Rd. Moung Nakhon Ratchasima 30000, Thailand (somphinith.mu@rmuti.ac.th)

[2]Rajamangala University of Technology Isan, Faculty of Sciences and Liberal Arts, 744 Suranaray Rd. Moung Nakhon Ratchasima 30000, Thailand

Keywords: Fenton oxidation, Heterogeneous catalyst, Nakhon Ratchasima Province, Thailand

Abstract

Textile is very famous and it is one of the major industries in Thailand. Each year, dye factories use a lot of dye and release it into the natural water resources such as Pakthongchai's silk village, Nakhon Ratchasima Province, Thailand. The dye components cause water pollution and they are hardly subject to degradation as recalcitrant chemicals. One way to remove Methyl Orange (MO) in dye adsorption process is using zeolite adsorbents. The pottery industry is famous and important like the textile industry in Thailand. Waste products from Dankwian's pottery village, Nakhon Ratchasima Province, Thailand, have a lot of Clayton-iron used for the adsorption process of dye wastewater treatment. This study used the component of clayton-iron for dye adsorption process to treat dyeing water. The experimentally obtained results show that the defects of uncoated pottery are capable of catalyzing the Fenton process. At the best conditions (at 60 mg/L methyl orange, 50 mL of the wastewater, 10 mg/L pottery scraps, 10 mM hydrogen peroxide, pH 3, and 90 min reaction time at room temperature, the methyl orange disposal efficiency reached 90%.

1 Introduction

Fenton oxidation process is widely used for wastewater treatment in many industries to decompose hardly biodegradable organics because it is easy to be applied and it has low cost. This process uses ferrous ions solution to interact with hydrogen peroxide (H_2O_2) in order to obtain hydroxyl radicals, which have a high potential to oxidize [1, 2]. Although its interaction rate is fast, it is difficult to separate ferrous ions as catalyst from wastewater. Moreover, in some cases, the combination of substances in the wastewater and ferrous ions form high-stable complex substances that cause a reduction of catalyst's ability. In order to get rid of this problem, extensive research efforts have been spent to study the use of heterogeneous catalysts in the Fenton oxidation process. Nowadays, catalysts used in this process are often heterogeneous catalysts such as iron minerals [3], crystals of iron oxide on nano-composites

[4], iron on alumina [5], iron on zeolite [6], iron on clay minerals developed to be pillared clays [7], iron exchange resin [8], and iron on activated carbon [9]. An iron containing catalyst will react with hydrogen peroxide to form hydroxyl radicals [1] used in the decomposition of hardly biodegradable organics, for example, organic dyes [10, 11], antibiotics [12], pesticides [13-15], landfill leachates [16, 17], explosives [18], and phenols [19, 20] because of its high redox potential.

Silk produced in Pakthongchai District, Nakhon Ratchasima Province, Thailand is very famous and it is used to decorate houses, to make clothes and to be exported to many countries. It is produced in both household handcraft and industry. Production process of silk in this district is modern; for instance, they use thread that is made in a factory and dye it with chemical colorant. Furthermore, silk is weaved by modern machines. Weaved silk has a beautiful and modern pattern and meets the requirements of consumers. However, the process of dyeing silk uses chemicals, which produces a huge amount of wastewater each day.

Pottery in Dankwian Sub-district, Chokchai District, Nakhon Ratchasima Province, Thailand is earthenware type, which has been developed from household handcraft to industry in order to distribute in and outside the country. From past to present, production processes, trading system and product styles have been developed. Nevertheless, the higher the pottery product is popular, the higher the quantity of defects and scraps are.

Due to the initial study, soil used to make pottery in Dankwian Sub-district, Chokchai District, Nakhon Ratchasima Province is sedimentary soil heaped naturally by Moon River. The soil has a large amount of oxide iron [3-9]. The aim of this research was to treat wastewater from the silk factory by using the defects of uncoated-pottery as Fenton catalyst since they are easy to purchase and eco-friendly. In this research, methyl orange (MO) was representing dyeing color of azo type.

2 Material and Methods

In this research, the property of the Fenton catalysis of iron ore in a defect of uncoated-pottery is examined by using an MO solution in distilled water as model, which is used as dyeing color in silk factory. Moreover, the effects of pH, the quantity of catalyst, and the concentration of hydrogen peroxide were investigated, as detailed below.

1. Set the pH of MO solution at 3, 5 and 7,
2. Use the quantity of catalyst at 0.05, 0.2, 0.5 and 1.0 g/50 mL of MO (or 1, 4, 10 and 20g/L, respectively),
3. Use 30% of hydrogen peroxide at 0.05, 0.1, 0.2 and 0.5 mL (or the concentration of hydrogen peroxide at 6, 10, 19, 39 and 97 mM, respectively),
4. Use the concentration of MO at 60, 100, 150 and 200 mg/L, and
5. Adjust the temperature at 25, 30 and 50 °C

All experiments were carried out in 125 mL-Erlenmeyer flasks with 50 mL of 60 mg/L MO. The pH of the solutions was adjusted to the desired values by using 3M H_2SO_4 and after that the uncoated pottery scraps (UPS) were added. The reactions were initiated by adding a predetermined amount of H_2O_2 solution to the flask. The mixture solution was stirred with a magnetic stirrer. After the study period, the reaction mixture was centrifuged to remove the catalyst. The concentrations of the MO were measured using a double beam UV-vis spectrophotometer at 510 nm. The decolorization efficiency of MO was calculated using the following equation:

$$\text{Decolorization Efficiency (\%)} = \frac{C_0 - C_t}{C_0} \times 100$$

where C_0 (mg/L) is the initial concentration of MO and C_t (mg/L) is the concentration of MO at the time, t (min).

3 Results and Discussion

3.1 Effect of pH on methyl orange degradation

The effect of pH on the degradation of MO was tested at pH 3, 5 and 7. The experimental results show that pH 3 gives the highest percentage of MO removal (see Figure 1). Furthermore, more than 90% within 90 min can be disposed of. This result is in accordance with the research of [21]. They found that Fenton reaction will be performed best at pH 2.5-3.5 because hydrogen peroxide and ferrous ion are stable under acidic conditions. By this, Fenton oxidation is efficient to decompose MO. In addition, MO structure in azo position changes to a single bond, which is weaker than double bond because protons react at the double bond position. So, the decomposition occurs easier on the account of reaction between hydroxyl radicals formed from Fenton catalyzing and single bond. On the other hand, hydrogen peroxide is unstable under alkaline conditions. Oxygen (O_2) and water (H_2O) are easily formed through hydrogen peroxide decomposition, and the oxidization efficiency of the organics reduces [22]. Furthermore, the structure of MO at the azo position under this alkaline condition is a double bond at nitrogen, which is a strong bond. Hence, it is quite difficult for hydroxyl radicals to react with MO. As a result, the percentage of MO removal was less.

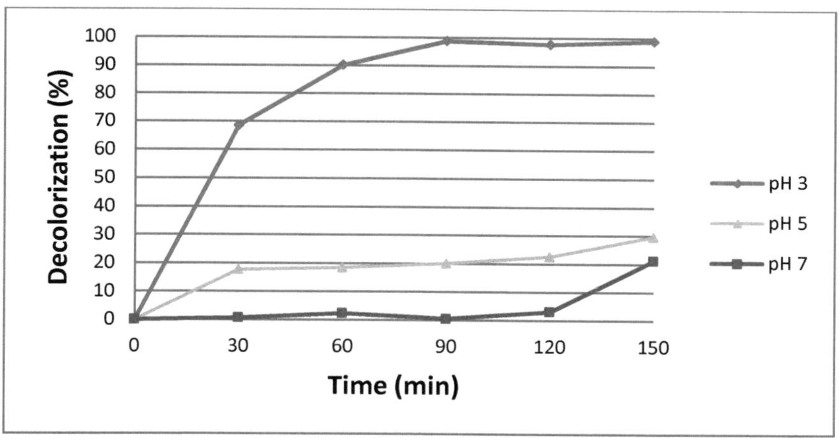

Figure 1: The pH influence on the percentage of the decomposition of methyl orange

3.2 Influence of the catalyst amount on the MO degradation
The experimentally results obtained of using the quantities of catalyst at 0.05, 0.2, 0.5 and 1.0 g/50 mL of methyl orange (or 1, 4, 10 and 20 g/L, respectively), showed that the more the quantity of catalyst increased, the more the efficiency of MO disposal increased, as well. Increasing the quantity of catalyst accelerates the decomposition of hydrogen peroxide (Figure 2). Subsequently, the quantity of hydrogen radicals formed gets higher, and so MO decomposition increased also [23].

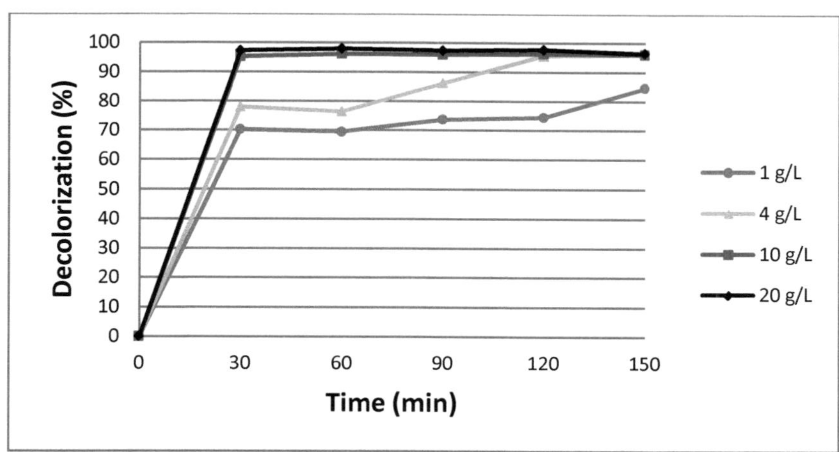

Figure 2: The influence of the catalyst quantity on the percentage of the MO degradation

3.3 Influence of the hydrogen peroxide amount on the MO degradation
The results from using 30% of hydrogen peroxide at 0.05, 0.1, 0.2 and 0.5 mL (or the concentration of hydrogen peroxide at 6, 10, 19, 39 and 97 mM, respectively), showed that the MO

removal percentage was the lowest at 6 mM of hydrogen peroxide, because hydroxyl radicals formed from hydrogen peroxide decomposition was not sufficient enough for the Fenton reaction. Then, dyeing color removal rate increased promptly when hydrogen peroxide concentration was set at 10, 19 and 39 mM in 30-60 min. But then, the dyeing color removal rate at different concentrations became similar during 90-150 min (Figure 3).

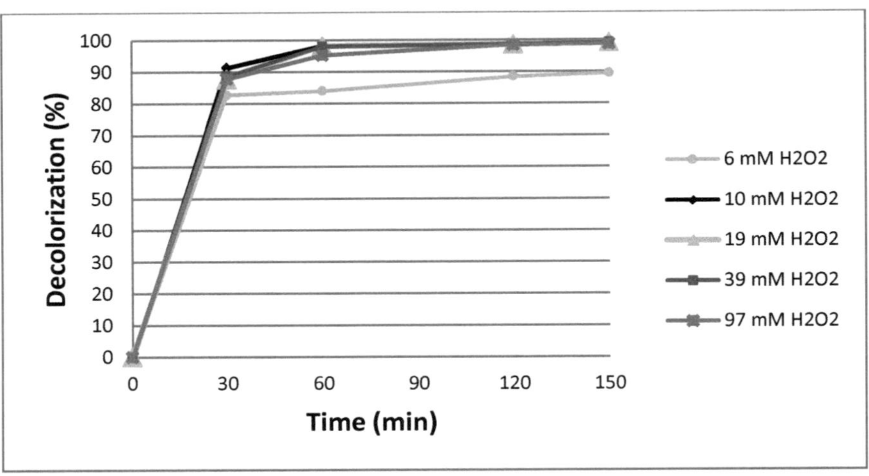

Figure 3: The influence of 30% hydrogen peroxide on the percentage of MO degradation

Increasing hydrogen peroxide concentration made the quantity of hydroxyl radicals becoming higher. Due to increasing concentration of the hydroxyl radicals, the efficiency of MO removal tended to increase. However, removal rate was reduced during 30-60 min when hydrogen peroxide concentration was set over the proper concentration of 97 mM. Here, hydroxyl radical could react with hydrogen peroxide to give perhydroxyl radical, which is an oxidizing agent with a lower redox potential compared with hydroxyl radical itself [24, 25].

3.4 Effect of MO concentration on its degradation rate
According to the experimental results obtained, at 200 mg/L of MO, the pottery scraps can be used to dispose of the dyestuff. The results also showed that, when the concentration of MO increased, the percentage of MO degradation increased, as well. Due to a short life time of hydroxyl radicals (approximately 20 µs), they are able to react with organics immediately. Hence, increasing the number of MO molecules increased the probability of the collision between organics and the oxidizing substance as well, which caused a higher efficiency of MO disposal [26].

3.5 Effect of the reaction temperature on MO degradation rate
Experimental results showed that the more temperature increased, the more the efficiency of MO degradation increased, as well; because increasing temperature accelerated the formation

of the hydroxyl radicals, making MO molecules having enough energy over activation energy [27], and made iron species getting a higher valence [23].

4 Conclusions

In order to treat wastewater in silk industry in Pakthongchai District, Nakhon Ratchasima Province, Thailand, the defected uncoated-pottery from Dankwian Sub-district, Chokchai District, Nakhon Ratchasima Province are used as Fenton catalyst of methyl orange solution. The variables in Fenton reaction such as pH, the quantity of catalyst, the concentration of methyl orange and hydrogen peroxide, temperature and reaction time were investigated in this research project. The experimental results obtained showed that the defects of uncoated pottery are capable of catalyzing the Fenton process. At the best conditions (60 mg/L of methyl orange, 50 mL of methyl orange, 10 mg/L of pottery scraps, 10 mM of hydrogen peroxide, pH 3 and 90 min of reaction time at room temperature, the methyl orange disposal efficiency reached 90%. The results of this investigation suggest that it is possible to treat the dyeing wastewater with another waste from the pottery industry, and to make an efficient use of the recombination of two waste streams to the benefit of the environment.

5 Acknowledgements

The authors wish to extend their sincere appreciation to DAAD and Exceed Swindon project for providing financial support to participate at the expert workshop on "Sustainable Waste Management in Developing Countries and Emerging Economies". They also sincerely thank Faculty of Engineering, Rajamangala University of Technology Isan for providing some financial assistance.

6 References

[1]   Haber, F., Weiss, J., The Catalytic decomposition of hydrogen peroxide by iron salts. Proceedings of the Royal Society A: Mathematical, Physical & Engineering Science, 1934, 147, 332-351.

[2]   Walling, C., Fenton's reagent revisited. Acc. Chem. Res., 1975, 8, 125-131.

[3]   Baldrian, P., Merhautova`, V., Gabriel, J., Nerud, F., Stopka, P., Hruby` M., Bernes, M.J., Decolorization of synthetic dyes by hydrogen peroxide with heterogeneous catalysis by mixed iron oxides, Appl. Catal., B. 2006, 66, 258-264.

[4]   Melero, J.A., Martinez, F., Botas, J.A., Molina, R., Pariente, M.I., Heterogeneous catalytic wet peroxide oxidation systems for the treatment of an industrial pharmaceutical waste-water. Water Res., 2009, 43, 4010-4018.

[5]   Al-Hayek, N., Dore, M., Oxidation of phenols in water by hydrogen peroxide on alumina supported iron. Water Res., 1990, 24, 973-982.

[6]   Kasiri, M.B., Aleboyah, H., Aleboyah, A., Degradation of Acid Blue 74 using Fe-ZSM5 zeolite as a heterogeneous photo-Fenton catalyst. Appl. Catal. B., 2003, 84, 9-15.

[7]   Mishra, T., Mohapatra, P., Parida, K.M., Synthesis, characterization and catalytic evaluation of iron–manganese mixed oxide pillared clay for VOC decomposition reaction. Appl. Catal. B., 2008, 79, 279-285.

[8]   Liou, R.M., Chen, S.H., Hung, M.Y., Hsu, C.S., Lai, J.Y., Fe (III) supported on resin as effective catalyst for the heterogeneous oxidation of phenol in aqueous solution. Chemosphere, 2005, 59, 117-125.

[9]   Zazo, J.A., Casas, J.A., Mohedano, A.F., Rodriguez, J.J., Catalytic wet peroxide oxidation of phenol with Fe/active carbon catalyst. Appl. Catal. B., 2006, 65, 261-268.

[10] Núñez, L., García-Hortal, J.A., Torrades, F., Study of kinetic parameters related to the decolourization and mineralization of reactive dyes from textile dyeing using Fenton and photo-Fenton processes. Dyes Pigm., 2007, 75, 647-652.

[11] Cheng, M., Song, W., Ma, W., Chen, C., Zhao, J., Lin, J., Zhu, H., Catalytic activity of iron species in layered clays for photodegradation of organic dyes under visible irradiation. Appl. Catal. B., 2008, 77, 355-363.

[12] Bobu, M., Yediler, A., Siminiceanu, I., Schulte-Hostede, S., Degradation studies of Ciprofloxacin on a pillared iron catalyst. Appl. Catal. B., 2008, 83, 15-23.

[13] Chan, K.H, Chu, W., Model applications and mechanism study on the degradation of atrazine by Fenton's system. J. Hazard. Mater., 2005, 118, 227-237.

[14] Barreiro, J.C., Capelato, M.D., Martin-Neto, L., Hansen, H.C.B., Oxidative decomposition of atrazine by a Fenton-like reaction in a H_2O_2/ferrihydrite system. Water Res., 2007, 41, 55-62.

[15] Oller, I., Malato, S., Sánchez-Pérez, J.A., Maldonado, M.L., Gassó, R., Detoxification of wastewater containing five common pesticides by solar AOPs-biological coupled system. Catal. Today, 2007, 129, 69-78.

[16] Deng, Y., "Physical and oxidative removal of organics during Fenton treatment of mature municipal landfill leachate". J. Hazard. Mater, 2007, 146, 334-340.

[17] Primo, O., Rueda, A., Rivero, M.J., Ortiz, I., An integrated process, Fenton reaction ultrafiltration, for the treatment of landfill leachate: pilot plant operation and analysis. Industrial and Engineering Chemistry Research, 2008, 47, 946-952.

[18] Liou, M.J., Lu, M.C., Catalytic degradation of explosives with goethite and hydrogen peroxide. J. Hazard. Mater., 2008, 151, 540-546.

[19] Araña, J., Pulido, M.P., Rodríguez, L.V.M., Peña, A.A., Doña, R.J.M., González, D.O., Pérez. P.J., Photocatalytic degradation of phenol and phenolic compounds part I. Adsorption and FTIR study. J. Hazard. Mater., 2007, 146, 520-528.

[20] El-Hamshary, H., El-Sigeny, S., Abou Taleb, M.F., El-Kelesh, N.A., Removal of phenolic compounds using (2-hydroxyethyl methacrylate/acrylamidopyridine) hydrogel prepared by gamma radiation. Sep. Purif. Technol., 2007, 57, 329-337.

[21] Chen, A., Ma, X., Sun, H., Decolorization of KN-R catalyzed by Fe-containing Y and ZSM-5 zeolites. J. Hazard. Mater., 2008, 156, 568-575.

[22] Jeong, J, and Yoon, J., pH effect on OH radical production in photo/ferrioxalate system. Water Res., 2005, 39, 2893-2900.

[23] Hassan, H., Wan, Z., Fenton-like oxidation of Acid Red 1 solution using heterogeneous catalyst based on ball clay. Intern. J. Environ. Sci. Develop., 2011, 2(3), 218.

[24] Bautista, P., Mohedano, A.F., Gilarranz, M.A. Casas, J.A., Rodriguez, J.J., Application of Fenton oxidation to cosmetic wastewater treatment. J. Hazard. Mater., 2006, 143, 112-120.

[25] Neyens, E., Baeyens, J., A review of classic Fenton's peroxidation as an advanced oxidation technique. J. Hazard. Mater., 2003, 98, 33-50.

[26] Hassan, H., Hameed, B.H., Decolorization of Acid Red 1 by Fenton-like using acid-activated clay. IEEE Symposium Humanities, Science and Engineering Research 2012, pp. 287-292.

[27] Xu, H.-Y., Prasad, M., Wang, P., Enhanced Removal of Phenol from Aquatic Solution in a Schorl-catalyzed Fenton-like System by Acid-modified Schorl. Bull. Korean Chem. Soc., 2010, 31, 803-807.

ACRYLAMIDE DEGRADATION BY DIRECT PHOTOLYSIS (UV) AND CONJUGATED PHOTOLYSIS (UV/H$_2$O$_2$)

[1]Elvis Carissimi, [1]Jessica Martini, [1]Suzan Costa Zilli, [2]Tânia Mara Pizzolato

[1] Laboratory of Engineering for the Environment, Federal University of Santa Maria (UFSM), Av. Roraima, 1000, Santa Maria/RS, Brazil (ecarissimi@gmail.com)

[2] Laboratory of Analytical and Environmental Chemistry, Federal University of Rio Grande do Sul (UFRGS), Av. Bento Gonçalves, 9500, Porto Alegre/RS, Brazil

Keywords: acrylamide; advanced oxidation process; degradation kinetics; high performance liquid chromatography; UV light.

Abstract

Acrylamide (AA) is widely used in the production of polyacrylamide, which is used as a polymer for drinking water clarification and wastewater treatment. The use as a polymer is the main source of contamination of acrylamide in water for human consumption. WHO established a maximum of 0.5 µg/L residual level due its carcinogenic effects. Thus, the main goal of this work was to investigate the efficiency of UV and UV/H$_2$O$_2$ processes for the acrylamide degradation in aqueous solution. The experiments were carried out in a batch quartz photoreactor with a mercury vapor lamp inside a timber box to block radiation. Direct UV photolysis experiments were carried out (10 min interval of sampling) and UV/H$_2$O$_2$ conjugated photolysis (5 min interval of sampling). Monitoring of acrylamide degradation was carried out with a HPLC-UV. Results showed that AA has been degraded when employed UV and UV/H$_2$O$_2$ photolysis systems. The UV/H$_2$O$_2$ process resulted in 89% reduction in AA concentration almost instantaneously in less than 2 min, probably due to the generation of large amounts of hydroxyl radicals, since these are generated when H$_2$O$_2$ is exposed to UV. The use of UV radiation also resulted in a complete degradation of AA, but with a 360 min exposure.

1 Introduction

Acrylamide (AA) is a white crystalline solid with a C$_3$H$_5$NO molecular formula and with the molecular structure shown in Figure 1. Most of acrylamide produced is used as monomer in the production of polyacrylamide, which is used as a flocculant for water clarification for human consumption, and industrial wastewater and sanitary treatment [1]. Both substances, either in monomeric or polymeric form, were included in the chemicals list of the Environmental Protection Agency of the United States [2] to be screened and evaluated with respect to its endocrine disruptor potential. World Health Organization [1] and the Brazilian Potable Water Standards [3] established a maximum of 0.5 µg/L residual level due its carcinogenic effects. Despite the severe legislation about the low concentration residual of acrylamide allowed to remain in the potable water, it is not a parameter evaluated in water treatment facilities.

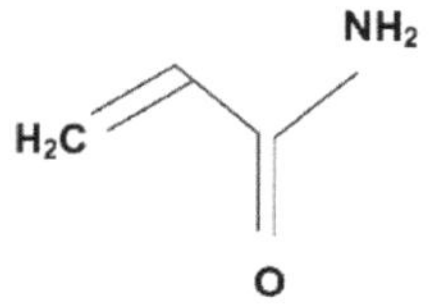

Figure 1. Acrylamide structural formula (C_3H_5NO)

The most important source of acrylamide contamination of water for human consumption is the use of polyacrylamide as a polymer for water treatment, which contains residual levels of monomer [1, 4]. In conventional plants, the polymer enables the water clarification process as for the purpose of absorption and formation of intermolecular bridges with suspended particulates (polymer bridge flocculation), it is possible to form larger aggregates which are more easily separated from the medium [5, 6].

Thus, in systems that use acrylamide in the flocculation step, a residual contaminant may result from the water treatment process. According to the World Health Organization, in general, the maximum concentration authorized polymer is 1 mg/L. Considering that the monomer content is about 0.05%, the theoretical maximum monomer concentration in the treated water is 0.5 µg/L [1]. Because of this, it is recommended as maximum allowed concentration of 0.5 µg/L acrylamide in drinking water. The risk of cancer associated with such concentration is 10^{-5}, which is a case of cancer per 100,000 inhabitants ingest water with the reference concentration for a period of 70 years [1]. Brazilian law follows the recommendation of WHO, as determined by Ordinance of the Ministry of Brazilian Health No. 2914/2011 [2].

Research to evaluate the effects of acrylamide in laboratory rats have been held since the 1980s, with results that prove its toxicity, especially in brain function with interference with neurotransmitters [7, 8, 9]. At the same time, other researchers have shown in the laboratory the carcinogenic effect of the substance [10, 11]. While the main concern at that time be exposure to acrylamide workers during polymer production or construction (with its use for waterproofing structures), it was hypothesized that the polyacrylamide used in water treatment could also pose risks to human health when consuming the treated water [10]. In humans, it is known that the metabolite of acrylamide (glycinamide) forms a reactive epoxide with DNA which confers genotoxic potential [12]. Treatment methods for disposal of acrylamide water or sewage are not employed, and there is no record of studies investigating the removal or degradation of it. In this sense, the advanced oxidation processes (AOP) became a promising technology for treating wastewater containing organic compounds of difficult biodegradability [13, 14]. Such processes can act on AA and turn it into less dangerous compounds or even mineralize it completely.

AOP represent a group of chemical-oxidative processes characterized by the generation of hydroxyl radicals ($HO^{\bullet}$), which are stronger oxidizers able to oxidize and mineralize almost all organic molecules [15]. Such technologies are even recommended by the American Association of

Work on Water as the most effective techniques for removing trace concentrations of endocrine disrupters [16]. Therefore, the goal of this work was to investigate the efficiency of direct photolysis processes (UV) and combined with hydrogen peroxide (UV / H_2O_2) in the acrylamide degradation in aqueous solution.

2 Experimental

2.1 Chemicals

Acrylamide (99%), catalase (bovine liver) (Sigma-Aldrich) and PA hydrogen peroxide 30% (Merck) were used in the degradation processes. Ultrapure water used and the mobile phase was obtained by Milli-Q system (Full 5). Acetonitrile and HPLC grade methanol (Sigma-Aldrich) was also employed as mobile phase to the chromatographic analysis.

AA degradation and monitoring via HPLC / UV

The monitoring of the degradation of acrylamide was carried out via determination of the chromatographic peak area in a high performance liquid chromatograph (HPLC) Shimadzu (LC-20 AD model) coupled to the detector UV-VIS (SPD-20A model) in the Laboratory of Analytical Chemistry and Environmental Institute of Chemistry, UFRGS (LQAA-IQ-UFRGS). Chromatographic separation occurred in a C18 column (250 mm x 4.6 mm; 5 µm) (Thermo Scientific) with mobile phase composed of (a) water (60%) and (b) acetonitrile (40%) in isocratic mode with a 0.8 mL min^{-1} flow rate. The injection volume was 20 µL, and the total time of analysis was 7 min at a wavelength of 210 nm. Prior to injection, samples were filtered through syringe filter PTFE 0.45 µm (Vertical Chromatography). Catalase was used to decompose the residual hydrogen peroxide when needed.

Degradation studies

The experiments were carried out in batch mode in a 225 mL photoreactor in the Laboratory of Analytical and Environmental Chemistry at UFRGS. The UV light was provided by an 80W mercury vapour lamp without the outer bulb. The system temperature was maintained by a flow of tap water surrounding the lamp. A magnetic stirrer was responsible for maintaining the solution under stirring. Samples were collected by syringe and the temperature monitored by a digital thermometer skewer (Incoterms). The system was kept in a timber box to block radiation. Direct photolysis experiments were carried out with photolysis (UV) and conjugated photolysis (UV / H_2O_2).

AA was used in ultrapure water solution at 5 mg/L concentration. The good sensitivity of the HPLC-UV method allows this reduction in concentration while maintaining good detection. For each experiment, the time for sampling to monitor the degradation was defined. Samples for AA analysis via HPLC-UV contained about 1.5 mL. The same procedures were kept for all assemblies: after lamp ignition, 5 minutes were waited for the initial heating, connected to cooling water and stirring. At the beginning and at the end of each batch, pH of the solution was measured (pH meter Digimed MD-22).

The experiments of direct photolysis (UV) were carried out in three batches for minimum interference of the volume change during the process and collection of rates. Where: (a) first set:

sampling at 0; 10; 20; 30 and 60 min; (b) second set: sampling at 0; 120; 180; 240 and 300 min; (c) third set: sampling at 0; 360; 420; 480 and 600 min.

For conjugate photolysis studies (UV / H_2O_2), the organization of the collection were as follows: (a) first set: 0; 2; 5; and 10 min; (b) second set: 0; 10; 12; 14; 20 and 30 min. In these studies 0.5 mL of H_2O_2 30% were injected into the system at the beginning of the process (time 0). All studies were carried out in triplicates.

3 Results and Discussion

3.1 AA degradation by conjugated photolysis (UV / H_2O_2)

The use of the conjugated process (UV/H_2O_2) resulted in a practically instantaneous degradation of the acrylamide. The sample at the time of 2 min already presented an AA concentration of 0.5 mg/L; i.e, a reduction of 89% (Figure 2). Due to this high rate of degradation it was not possible to determine the reaction kinetics. The pH of the solution decreased from pH 7.3 at the beginning to pH 4.8 after 3 min, probably because of the dissociation of H_2O_2, forming H^+ ions.

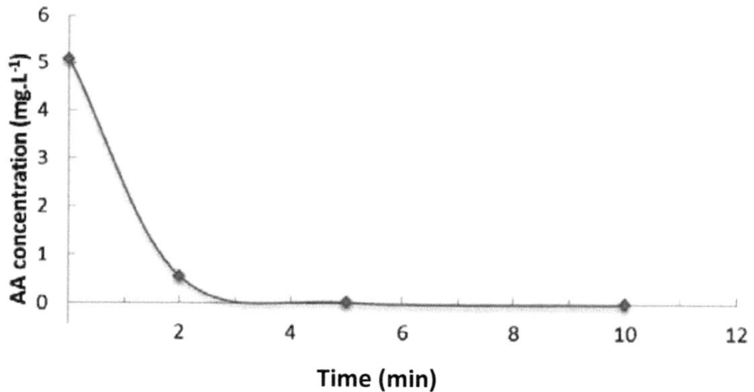

Figure 2:   AA degradation in aqueous solution over time, employing the UV/H_2O_2 process. (Sampling at 0, 2, 5 and 10 min)

The hydrogen peroxide in the presence of UV radiation generates hydroxyl radicals [17, 18]; the greater the concentration, the greater is the generation of hydroxyl radical. Thus, the rapid degradation of AA can be attributed to the generation of large amounts of hydroxyl radical shown in Equation (1). As the radical $HO^\bullet$ is a non-selective oxidant with a high oxidation power in complex molecules, the performance of this on AA molecule has been very effectively [15, 19, 20].

$$H_2O_2 + hv \rightarrow 2HO^\bullet \tag{1}$$

3.2 AA degradation by direct photolysis (UV)

The use of UV radiation also resulted in a complete degradation of AA within 360 min of exposure. Results relating concentration and time show a linearly decay on AA concentration with $R^2=0.9964$, featuring a degradation kinetics of zero order (Figure 3). Thus, is the kinetic constant is k= 0.0143 mg/L·min and consequently a half-life of 175 minutes.

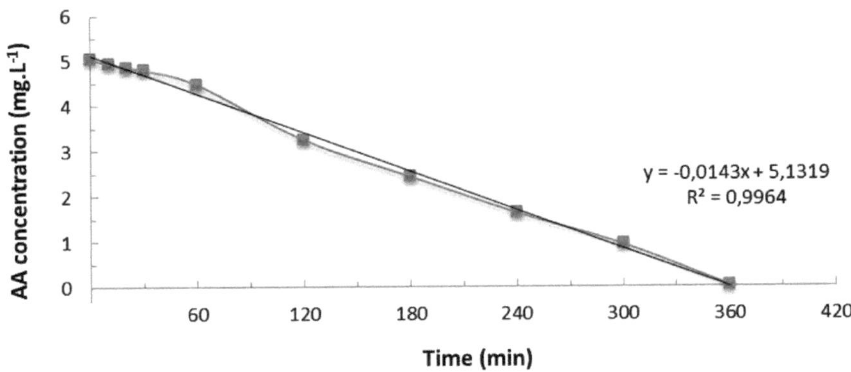

Figure 3: Zero-order kinetics for the degradation of acrylamide in aqueous solution in the presence of UV radiation. (Irradiation time: 0, 10, 20, 30, 60, 120, 180, 240, 300 and 360 min). R² = 0.9964 Trendline

From the AA solution in the presence of UV radiation and dissolved oxygen (H_2O + AA + hv + O_2) two sets of possible reactions between these may occur, resulting in the formation of hydroxyl radicals: the dissolved oxygen when irradiated (λ = 185 nm) produces H_2O_2 by means of ozone (Equation 2-4) [17]:

$$O_2 + hv \rightarrow 2O \tag{2}$$
$$O_2 + O \rightarrow O_3 \tag{3}$$
$$O_3 + H_2O \rightarrow H_2O_2 + O_2 \tag{4}$$

Yet it is possible to infer that radiation excites the AA molecule, as a single organic molecule through starting a chain reaction involving the formation of peroxide radical ($O_2^{-\bullet}$) to its dispro-portionation and formation of hydrogen peroxide (Equation 5-7 [21]):

$$AA + hv \rightarrow AA^* + e^- \tag{5}$$
$$O_2 + e^- \rightarrow O_2^{-\bullet} \tag{6}$$
$$2O_2^{-\bullet} + 2H^+ \rightarrow H_2O_2 + O_2 \tag{7}$$

After formation of hydrogen peroxide, the generation of the hydroxyl radicals may occur in the following ways: by photolysis (Equation 8) [22]; reduction (Equation 9); spontaneous dissociation in an alkaline medium (pH> 5.0) with the subsequent chain reactions for the formation of peroxide (Equation 10-12) [17]; and peroxide reaction with ozone (Equation 13 and 14) or with H_2O_2 to form the hydroxyl radical (15):

$$H_2O_2 + hv \rightarrow 2HO^\bullet \tag{8}$$

$$H_2O_2 + e^- \rightarrow HO^{\bullet} + OH^- \tag{9}$$

$$H_2O_2 \rightleftharpoons HO_2^- + H^+ \tag{10}$$

$$HO_2^- + O_3 \rightarrow O_3^{-\bullet} + HO_2^{\bullet} \tag{11}$$

$$HO_2^{\bullet} \rightleftharpoons O_2^{-\bullet} + H^+ \tag{12}$$

$$O_3 + O_2^{-\bullet} \rightarrow O_3^{-\bullet} + O_2 \tag{13}$$

$$O_3^{-\bullet} + H^+ \rightarrow HO^{\bullet} + O_2 \tag{14}$$

$$H_2O_2 + O_2^{-\bullet} \rightarrow HO^{\bullet} + OH^- + O_2 \tag{15}$$

Due to the complexity of the reaction set, responsible for the degradation of AA initiated by UV radiation, it is very difficult to establish the real degradation of the medium. However, it is seen that for the photolysis process, the presence of dissolved molecular oxygen (O_2) for generating the more active oxidizing AOP is indispensable for the formation of hydroxyl radicals [22].

In the case of bench reactor used in the test, while O_2 was consumed by the reactions, the intense contact of the medium with the atmosphere and the stirring was responsible for the natural addition of O_2 into the solution, following the equilibrium described by Henry's Law [23], as occurs in natural systems.

Processes using UV and UV / H_2O_2 are employed for the degradation of many compounds present in domestic and industrial effluents which generally are not removed in conventional treatment systems. One can cite the case of florfenicol antibiotics and thiamphenicol [18], drugs [24, 25, 26, 27], biocides and pesticides [28], surfactants [29], phthalates [30], dyes [31], among others. The great advantage of direct photolysis or combined processes (including the use of catalysts) is the possibility to use solar radiation as a source of UV radiation.

4 Conclusions

Acrylamide was degraded by direct photolysis conjugated system and photolysis. At first, presented a rate constant k = 0.0143 mg/L min^{-1}, being completely degraded after 360 min. Then, a 2 min reaction time was sufficient to degrade 89.3% of the AA dissolved, and being completely degraded after 5 min. The combined process also resulted in the formation of unidentified by-products. Photolysis processes in various configurations for water and sewage treatment are responsible for the degradation of various contaminants, including endocrine disrupters. Given the various acrylamide contamination sources and the risks associated with water consumption with the same, the improvement of these techniques and their incorporation into the plants for water and wastewater treatment would be of great importance for the health of the population.

5 Acknowledgements

Authors thank to DAAD and Exceed Swindon for the financial support to attend the Expert Workshop on *"Wastewater Treatment and Reuse for Metropolitan Areas and Small Cities"* in Recife (Brazil) September 11-17, 2016. Authors also thank to FAPERGS (Project n. 12/2291-8) for the financial support and CAPES for the Suzan Costa Zilli master scholarship.

6 References

[1]   World Health Organization. Guidelines for Drinking-water Quality, 2011. Available online at: www.who.int/water_sanitation_health/publications/2011/dwq_guidelines/en/

[2]   US.EPA. Universe of Chemicals. 2012. Available online at: www.epa.gov/endo/pubs/edsp_chemical_universe_list_11_12.pdf

[3]   BRASIL. MINISTÉRIO DA SAÚDE. Portaria nº 2914 de 12 de dezembro de 2011.

[4]   Dearfield, K., Douglas, G.R., Ehling, U.H., Moore, M.M., Sega, G.A., Brusick, D.J., Acrylamide: A review of its genotoxicity and an assessment of heritable genetic risk. Mut. Res. 1995, 330, 71-99.

[5]   Biggs, S., Habgood, M., Jameson, G.J., Yan, Y., Aggregate structures formed via a bridging flocculation mechanism. Chem. Eng. J. 2000, 80, 13-22.

[6]   Carissimi, E., Rubio, J., Polymer-bridging flocculation performance using turbulent pipe flow. Min. Eng. 2015, 70, 20-25.

[7]   Bond, S.C., Tilson, H.A., Agrawal, A.K., Neurotransmitter Receptors in Brain Regions of Acrylamide-Treated Rats II: Effects of Extended Exposure to Acrylamide. Pharmacol. Biochem. Beh. 1981, 14, 533-537.

[8]   Ali, S.F., Acrylamide-induced changes in the monoamines and their acid metabolites in different regions of the rat brain. Toxic. Lett. 1983, 17, 101-105.

[9]   Agrawal, A.K., Squibb, R.E., Bondy, S.C., The effects of acrylamide treatment upon the dopamine receptor. Toxicol. Appl. Pharmacol. 1981, 58, 89-99.

[10]   Bull. R.J., Carcinogenic and mutagenic properties of chemicals in drinking water. Sci. Total Environ. 1985, 47, 385-413.

[11]   Johnson, K.A., Gorzinski, S.J., Bodner, K.M., Campbell, R.A., Wolf, C.H., Friedman, M.A., Mast, R.W., Chronic toxicity and oncogenicity study on acrylamide incorporated in the drinking water of Fischer 344 rats. Toxicol. Appl. Pharmacol. 1986, 85, 154-68.

[12]   Soares, C.M.D., Determinação dos Teores de Acrilamida em Alimentos. Master Thesis, (Mestrado em Química Analítica Ambiental) Porto University, 2007. In Portuguese.

[13]   Ayoub, K., van Hullebusch, E., Cassir, M., Bermond, A., Application of advanced oxidation processes for TNT removal: A review. J. Hazard. Mat. 2010, 178, 10-28.

[14]   Bila, D.M., Dezotti, M., Desreguladores Endócrinos no Meio Ambiente: Efeitos e Consequências. Quim. Nova 2007, 30, 651-666. (In Portuguese)

[15]   Esplugas, S., Bila, D.M., Krause, L.G.T., Dezotti, M., Ozonation and advanced oxidation technologies to remove endocrine disrupting chemicals (EDCs) and pharmaceuticals and personal care products (PPCPs) in water effluents. J. Hazard. Mat. 2007, 149, 631–642.

[16]   AWWARF. American Water Works Association – Research Foundation. Removal of EDCs and Pharmaceuticals in Drinking and Reuse Treatment Processes. IWA Publishing: Denver, 2007.

[17]   Golimowski, J., Golimowska, K., UV-photooxidation as pretreatment step in inorganic analysis of environmental samples. Anal. Chim. Acta 1996, 325, 111-133.

[18]   Liu, N., Sijak, S., Zheng, M., Tang, L., Xu, G., Wu, M., Aquatic photolysis of florfenicol and thiamphenicol under direct UV irradiation, UV/ H_2O_2 and UV/Fe(II) processes. Chem. Eng. J. 2015, 260, 826-834.

[19]   Klavarioti, M., Mantzavinos, D., Kassinos, D., Removal of residual pharmaceuticals from aqueous systems by advanced oxidation processes. Environ. Int. 2009, 35, 402-417.

[20]   Orbeci, C., Untea, I., Nechifor, G., Segneanu, A.E., Craciun, M.E., Effect of a modified photo-Fenton procedure on the oxidative degradation of antibiotics in aqueous solutions. Sep. Purific. Technol. 2014, 122, 290–296.

[21]   Achterberg, E.P., van den Berg, C.M.G., In-line ultraviolet-digestion of natural water samples for trace metal determination using an automated voltammetric system. Anal. Chim. Acta 1994, 291, 213-232.

[22]   Luo, X., Zheng, Z., Greaves, J., Cooper, W., Song, W., Trimethoprim: Kinect and mechanistic considerations in photochemical environmental fate and AOP treatment. Water Res. 2012, 46, 1327-1236.

[23]   Baird, C., Química Ambiental. 2ª ed. Trad. M.A.L. Recio e L.C.M Carrera. Porto Alegre: Bookman, 2002.

[24]   Wols, B.A., Hofman-Caris, C.H.M., Harmsen, D.J.H., Beerendonk, E.F., Degradation of 40 selected pharmaceuticals by UV/H_2O_2. Water Res. 2013, 47, 5876-5888.

[25]   De La Cruz, N., Giménez, J., Esplugas, S., Grandjean, D., De Alencastro, L.F., Pulgarín, C., Degradation of 32 emergent contaminants by UV and neutral photo-fenton in domestic wastewater effluent previously treated by activated sludge. Water Res. 2012, 46, 1947-1957.

[26]   Arany, E., Szabó, R.K., Apáti, L., Alapi, T., Ilisz, I., Mazellier, P., Dombi, A., Gajda-Schrantz, K., Degradation of naproxen by UV, VUV photolysis and their combination. J. Hazard. Mat. 2013, 262, 151-157.

[27]   Donner, E., Kosjek, T., Qualmann, S., Kusk, K., Heath, E., Revitt, D., Ledin, A., Andersen, H., Ecotoxicity of carbamazepine and its UV photolysis transformation products. Sci. Total Environ. 2013, 443, 870-876.

[28]   De La Cruz, N., Dantas, R., Giménez, J., Esplugas, S., Photolysis and TiO_2 photocatalysis of the pharmaceutical propranolol: solar and artificial light. Appl. Catal. B: Environ. 2013, 130–131, 249-256.

[29]   Karci, A., Arslan-Alaton, I., Bekbolet, M., Ozhan, G., Alpertunga, B., H_2O_2/UV-C and Photo-Fenton treatment of a nonylphenol polyethoxylate in synthetic freshwater: Follow-up of degradation products, acute toxicity and genotoxicity. Chem. Eng. J. 2014, 241, 43–51.

[30]   Xu, B., Gao, N., Sun, X., Xia, S., Rui, M., Simonnot, M., Causserand, C., Zhao J., Photochemical degradation of diethyl phthalate with UV/H2O2. J. Hazard. Mat. 2007, 139, 132–139.

[31]   Timchak, E., Gitis, V., A combined degradation of dyes and inactivation of viruses by UV and UV/H_2O_2. Chem. Eng. J. 2012, 192, 164–70.

CIRCULAR ECONOMY
A NEW PERSPECTIVE FOR THE O&G INDUSTRY IN BRAZIL

[1]José Tavares Araruna Jr., [2]Patrício José Moreira Pires, [3]Ricardo Campos Vaqueiro, [1]Paola Machado Barreto Manhães

[1]Pontifical Catholic University of Rio de Janeiro, Department of Civil Engineering, Rua Marques de Sao Vicente 225, CEP 22451-900, Rio de Janeiro, Brazil, (araruna@puc-rio.br)
[2]Federal University of Espírito Santo, Brazil
[3]Petróleo Brasileiro S.A., Petrobras, Brazil

Keywords: brick manufacturing, circular economy, drilling cuttings, hazardous wastes

Abstract

This paper presents an innovative treatment technique for drilling cuttings, one of the most complex wastes of the oil industry. This treatment applies the concept of circular economy as it employs key parameters of its philosophy: the use of innovation through research of new materials employing recoverable resources by industrial symbioses. The treatment process involves the incorporation of drilling cuttings, up to 10% in weight, to the resulting mass used for manufacturing bricks. It consists of applying high temperatures, over 800 $^{\circ}$C, to the resulting mass in order to obtain a solid, resistant and non-hazardous material. In addition, its strength is higher than those from bricks made only by clayey mass. Its technological parameters (i.e., porosity, water absorption and unit weight) also show better values to those bricks made only with clayey materials. As a result, an environmental permit is issued by the State of Bahia Environmental Agency. So far, three independent brick making industries are now adding drilling cuttings to their process.

1 Introduction

Oil and gas (O&G) activities continue to grow across the globe. It is generally believed that O&G industry is mainly focused on profits with little concern for the environment, opposed to the concept of sustainability. However, demand for petroleum products continued to increase, and O&G continue to be the most widely used fuels. As energy demand continues to grow, petroleum companies must find and produce increasing quantities of O&G. But doing so requires more than merely ramping up production from traditional sources. As nearby, relatively easy-to-produce resources diminish, new sources and locations are being developed in Brazil. O&G exploration today requires the simultaneous consideration of a variety of economic, social, political, and environmental concerns [1, 2]. Historically, many foreign-operators started to invest in Brazil after the government opened the sector to competition. Increased drilling activity in Brazil both results in an increased amount of waste generated, air emission and water discharges. Once generated, managing these wastes in a manner that protects health and the environment will be essential for limiting operators' legal and financial liabilities. Therefore, operators will have to manage the waste properly and consistent with all relevant laws and regulations in order to overcome the challenges of the increase of activities in this dynamic industry.

An interesting alternative to overcome the aforementioned challenges is to apply the concept of circular economy on the waste management strategy. According to Stahel [3], circular economy is rather interesting since it closes loops in industrial ecosystems and minimize production of waste as can be seen in Figure 1. The author points out that circular economy models fall into two categories: (a) those that foster reuse and extend service life through repair, remanufacture, upgrades and retrofits, and (b) those that turn old goods into as new resources by recycling materials.

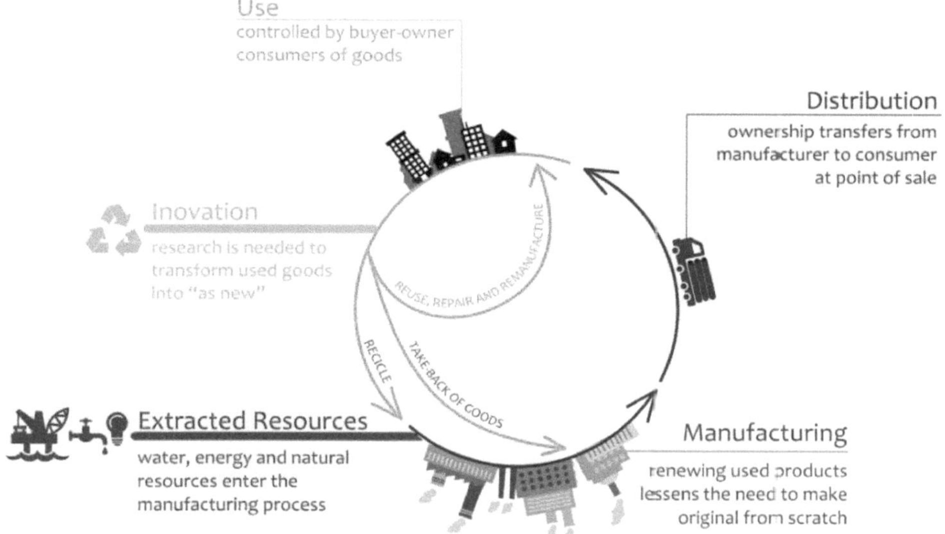

Figure 1: The concept of closing loops in circular economy (adapted from Stahel, 2016)

Waste Management Strategies in Exploration & Production Activities
Oil production in Northeast Brazil generates circa 20,000 tons of drilling cuttings each year. Drilling cuttings are considered the major source of wastes generated from the exploration and production activities. In Northeast Brazil, they consist mainly of a clayey soil mixed with n-paraffin, chloride and barite. They present high percentage of clay fraction, in which illite and kaolinite are the main clay minerals, and high plasticity. In general, drilling cuts are disposed of in dykes although they could be treated and, in some cases, reused [4, 5, 6]. According to Page et al. [7], the final destination of drilling cuttings is conditioned to the total hydrocarbon content, moisture content, salinity, and the existing clay fraction. Generally,

drilling cuttings are disposed of in dykes near the well base, are open and accumulates rainwater as seen in Figure 2 left. Dykes usually have a liner system unless the site possesses favorable geohydrology conditions [8].

Most of the drilling rigs that operate in Northeast Brazil employ centrifuges in order to reduce considerably the oil content of drilling cuttings before disposal. This strategy, shown in Figure 2 right, attains indexes below 8%. At the end of drilling operations, drilling muds are collected and treated adequately whereas the final destination of drilling cuttings is adopted considering their composition. If cuttings are not hazardous the current state of practice includes the placement of a vegetated final cover. If some constituent of the cuttings present a concentration slightly higher than the reference levels established by Brazilian legislation, cuttings might well be mixed with local soil to lower the concentration of this given parameter [9]. On the other hand, if the concentration of the hazardous constituent is well above the reference value, cuttings are removed from the dyke and disposed of in an industrial landfill [10] or treated adequately [11].

Figure 2: Drilling Cuttings Management Strategies: left: disposal on dykes and right: reduction of oil content using centrifuges

This paper seeks to provide an overview in how circular economy could turn wastes from the O&G industry into as new resources by recycling materials, showing the importance of research to attain this goal. An innovative treatment technique for drilling cuttings, one of the most complex wastes of the oil industry is developed. This waste is considered hazardous due to its high salt and hydrocarbon content, and poses serious threats to the environment. The treatment process involves the incorporation of drilling cuttings, up to 10% in weight, to the clayey soil mass used for manufacturing bricks. It consists on applying high temperatures, over 800 °C, to the resulting mass in order to obtain a solid, resistant and inert material. This treatment applies the concept of circular economy as it employs key parameters of its philosophy, the use of innovation through research of new materials employing recoverable resources by industrial symbioses.

2 Materials and Methods

Natural soils collected from the fields of Carmópolis (State of Sergipe), Anambé and Pilar (State of Alagoas), as well as drilling cuttings from E&P processes in both states were submitted to an experimental program in order to assess their physical, chemical and mineralogical properties. These properties were obtained by the use of the following Brazilian Standards (i.e., NBR-06457, NBR-06508, NBR-07181, NBR-06459 and NBR-07180) as well as the procedures described in EMBRAPA [12] and EMBRAPA [13]. Materials obtained from the borrow pits from Bandeira Brick Industry were named Cajueiro, materials obtained from INCELT Brick Industry were named Siriri, and drilling cuttings obtained from the drilling activities of oil wells CP129, CP1549, CP223D and CPANB03 have received their own denominations.

Ceramic pieces were manufactured employing the state of the practice in Brazil but using laboratory scale machines. Pieces resulted from the incorporation of four types of drilling cuttings on two types of clay materials as Table 1 shows. The properties from the materials obtained from the manufacturing process were assessed using the methods described in [14, 15, 16].

Table 1: Materials used in the experimental programme and the resulting mixes

Clayey Material	Drilling Cuttings	Resulting Mix
INCELT	CP129	INCELT + 5% of CP129
		INCELT + 10% of CP129
INCELT	CP1549	INCELT + 5% of CP1549
		INCELT + 10% of CP1549
BANDEIRA	CPANB03	BANDEIRA + 5% of CPANB03
		BANDEIRA + 10% of CPANB03
BANDEIRA	CP223D	BANDEIRA + 5% of CP223D
		BANDEIRA + 10% of CP223D

Measurements of pH of all materials were carried out using a pH sensitive electrode system (Thermo Scientific, model Orion Versa Star pH/ISE) in accordance to the procedure described in [12]. This measurement determines the degree of acidity of alkalinity in soil materials suspended in water and a 0.1 N potassium chloride solution. The mineralogical analyses of the materials were performed by X-Ray diffraction applying the powder method using a Siemens model D 5000 X-Ray diffraction apparatus [17]. Toxicity Characteristic Leaching Procedure was used to determine the mobility of organic and inorganic analytes present in the natural soils, in the drilling cuttings and in the resulting bricks [18].

Water absorption, porosity, density and flexural strength of bricks were determined applying the procedures described in [19]. These tests used a balance with a capacity of 210 g and sensitive to 0.001 g (MARTE, model AD200), watertight trays having an inside depth of 15 mm, a caliper having a scale ranging from 1 to 150 mm (Starret, model EC799A-E), a dial micrometer graduated to read in 0.01 mm (Starret, model 444.1MXRL-25), and a ventilated oven maintained at a temperature of 110 °C. Flexural strength was determined using a universal compression machine (MTS, model 810) employing a uniform speed of 1.0 mm/min.

3 Results and Discussion

All materials, natural clayey soils and drilling cuttings, present particle density ranging from 2.65 to 2.76 suggesting that their grains are constituted mainly from quartz and aluminium silicates. Natural soils present high percentage of particles with diameter less than 75 mm, an indicative of a high cation exchange capacity to adsorb ions presented in the drilling cuttings, as can be seen in Figure 3. Natural soils also present high plasticity suggesting that they may have 2:1 clay minerals.

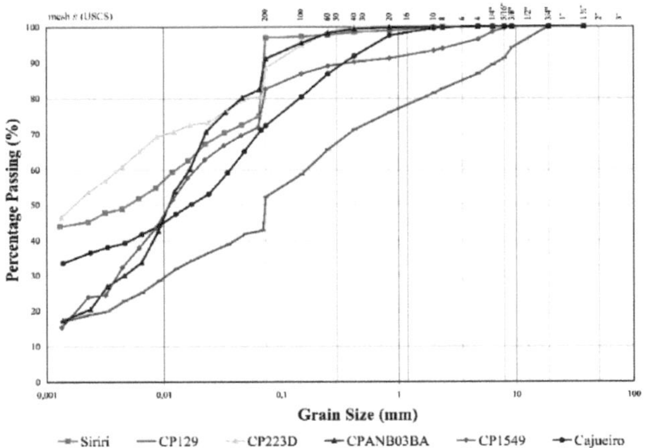

Figure 3: Grain distribution curve

Table 2 presents the results of pH measurements of all materials used in the experimental program. The results show that there is a negative difference between pH (soil/H_2O) and pH (soil/KCl) for Cajueiro and Siriri an indicative of the capacity of both soils in retaining cations. In practice, both soils might have the capacity to adsorb sodium present in the drilling cuttings.

Table 2: pH values

Material	pH (soil/H_2O)	pH (soil/KCl)
Cajueiro	7.3	5.5
Siriri	7.8	7.1
Drilling cuttings CP223D	8.0	7.9
Drilling cuttings CP1549	10.3	10,.
Drilling cuttings CP223D	8.9	8.,3
Drilling cuttings CPANB	10.0	10.0

Mineralogical analysis by X-Ray diffraction have shown that natural soils present quartz, kaolinite and mica as their main constituents while drilling cuttings present quartz, calcite and mica as their main constituents. Toxicity characteristic leaching procedure (TCLP) results were applied to classify the drilling cuttings as hazardous materials and natural soils as non-hazardous but inert materials.

The technological properties of the resulting bricks were assessed after the completion of the drying and calcination processes. As Figure 4 shows, a remarkable change in density as a result of increasing the amount of drilling cuttings on the mixing after the drying process was not observed. This was also the case for the calcination process.

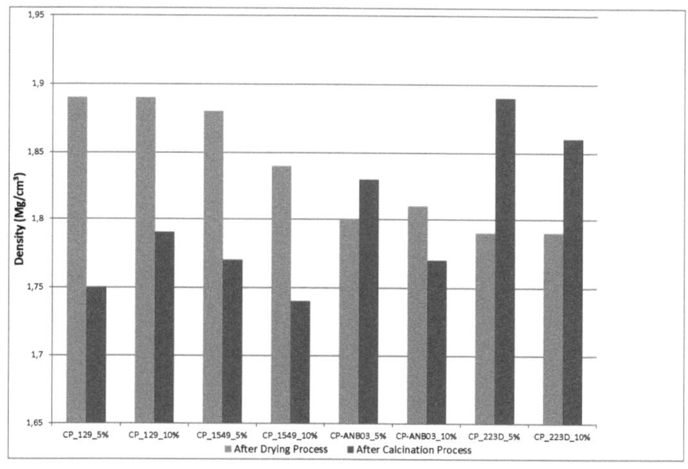

Figure 4: Density of bricks (*left bars*: after drying; *right bars*: after calcination)

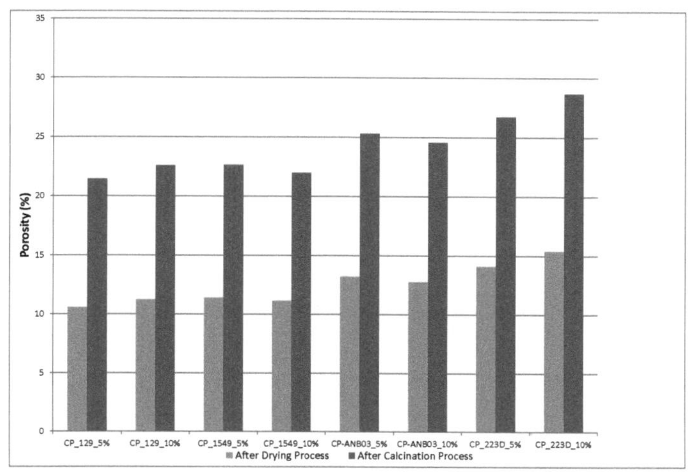

Figure 5: Porosity of bricks (*left bars*: after drying; *right bars*: after calcination)

In general, the results of tests of porosity, illustrated in Figure 5, revealed that the specimens made from INCELT presented a lower porosity value than the specimens made from BANDEIRA. However, the difference is quite small. It was also noted that the porosity of specimens has increased considerably as a result of burning.

Regarding water absorption, a pattern of variation of this parameter as a result of the increase in the level of incorporation of drilling gravel into the mix was not observed, as can be seen in Figure 6. In general, the results revealed that specimens made from ceramic material INCELT presented a lower water absorption value than specimens made from ceramic material BANDEIRA. This fact is due to the lower porosities observed in specimens made from ceramic material INCELT.

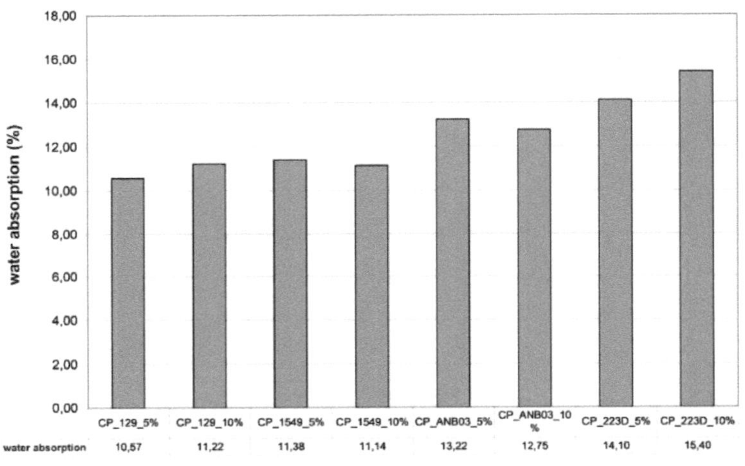

Figure 6: Water absorption

The ceramic pieces were subjected to three point tests in order to determine the ultimate flexural strength. In general, a reduction of the strength value with the increase of the levels of incorporation of drilling cuttings was found (Figure 7). Only the specimens made from the cuttings CP129 showed a slight increase in strength with the incorporation of drilling cuttings. However, the difference was small that for practical purposes it can be considered that there was no change in strength with the incorporation. The results presented in Figure 7 showed a great difference in strength values between the specimens made with INCELT material and those specimens made with BANDEIRA material. The difference in the values of porosity and water absorption already gave a clear indication of the difference that would be observed. The results also indicated that in addition to being more resistant, the pieces made with INCELT material mobilize maximum strength at a higher level of strain, i.e. the pieces made with INCELT material present a smaller stress-strain module.

Finally, the ceramic specimens resulting from the incorporation of drilling cuttings were classified in relation to their hazards by employing the Solid Waste Classification Brazilian Standard (NBR 10,004). Since levels of dissolved aluminum have supplanted the maximum value permitted by NBR 10,004 in its Annex G, the ceramic pieces made from the incorporation of CP129 and CP1549 drilling gravel to the SIRIRI soil were classified as non-hazardous but non-inert waste, getting the classification IIA, notwithstanding the level of incorporation.

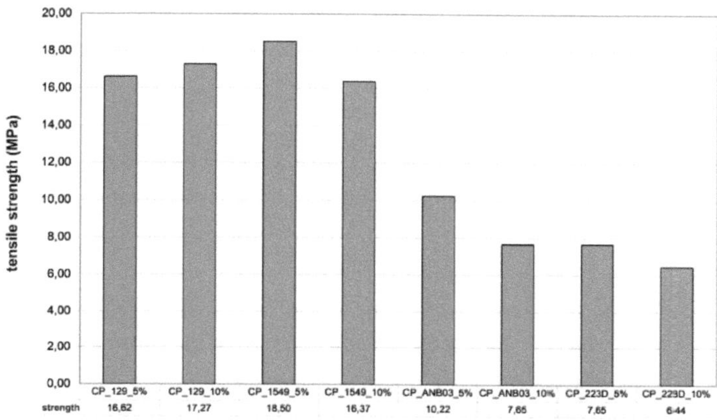

Figure 7: Average values from tensile strength

On the other hand, ceramic pieces obtained from the addition of 5% of the CPANB03 cuttings and 10% of the CP223B cuttings to CAJUEIRO soil were classified as non-hazardous but inert waste, getting the classification IIA. However, the pieces obtained from the incorporation of 10% of the CPANB03 cuttings and 5% gravel of the CP223B cuttings to the CAJUEIRO soil were classified as non-hazardous inert waste, getting the classification IIB.

4 Conclusions

The results obtained from the experimental program showed the feasibility of incorporating drilling cutting to natural soils to manufacture bricks. The resulting properties were considered excellent or at least acceptable when compared to the parameters established in Dondi [20]. Moreover, they were similar to those obtained from the most important centers of brick manufacturing in Brazil, i.e. Santa Gertrudes (SP) and Campos dos Goytacazes (RJ). It is also worth to mention that the tensile strength of the bricks obtained from the incorporation of drilling cuttings were at least two times higher than the values obtained from the incorporation of other wastes reported in the literature [21, 22, 23, 24].

The environmental hazards from the incorporation of drilling cuttings were assessed by TCLP tests. The results indicated that the bricks produced leach few inorganic constituents (e.g., barium, lead, fluoride and chromium). However, all of them are found at concentrations well below the maximum orientating values permitted by Brazilian Standards, suggesting that these were retained in the ceramic material. In relation to solubilized constituents, it was observed that in the vast majority of incorporations the dissolved aluminum concentrations presented was above the maximum allowed value. It should be noted that aluminum is one of the main constituents of clay soils as well as reservoirs rocks. The resulting bricks were classified as non-hazardous materials according to the Solid Waste Classification Brazilian Standard suggesting that they could well be used in the construction industry. In fact, the process was approved by the Environmental Agency of the State of Bahia and three industries are already using the incorporation in their manufacturing processes.

5 Acknowledgements

The authors wish to express their gratitude to DAAD and Exceed Swindon Project that made the participation at the event possible, and to Petrobras for its assistance and support.

6 References

[1]   OLF - Norwegian Oil Industry Association. Environmental report 2011 - The environmental efforts of the oil and gas industry Facts and trends (2011). Available at: http://www.olf.no/PageFiles/11829/Environmental%20Report%202011.pdf.

[2]   IPIECA -International Petroleum Industry Environmental Conservation Association; API - American Petroleum Institute; OGP - Oil and Gas Industry. Guidance on Voluntary Sustainability Reporting (2010). 2nd edition. 150p. Available at: http://www.api.org/ehs/performance/upload/voluntary_sustainability_reporting_guidance_2010.pdf. 2010.

[3]   Stahel, W.R., Circular Economy, *Nature*, 2016, 531, 435-438.

[4]   Drilling Waste Management Information System. *Fact Sheet – Beneficial Reuse of Drilling Wastes (2005)*. Available at http://www.web.ead.anl.gov/dwm/techdesc/reuse/index.cfm.

[5]   Al-Ansary, M., Al-Tabbaa, A., Stabilization/solidification of synthetic petroleum drill cuttings. *J. Hazard. Mater.* 2007, 141, 410-421.

[6]   Harbottle, M.J., Al-Tabbaa, A., Evans, C.W., A comparison of the technical sustainability of in situ stabilisation/solidification with disposal to landfill. *J. Hazard. Mater.* 141, 430-440.

[7]   Page, P.W., Greaves, C., Lawson, R., Boyle, F. *Options for the recycling of drill cuttings*, SPE 805583, Proceedings of the SPE/EPA/DOE Exploration and Production Environmental Conference (2003), San Antonio, TX.

[8]   Drilling Waste Management Information System. *Fact Sheet – Land Application* (2005b), available at http://www.web.ead.anl.gov/dwm/techdesc/land/index.cfm

[9]   Bansai, K.M., Sugiarto. *Exploration and Production Operations – Waste Management a Comparative Overview: US and Indonesia Cases*, SPE 54345, SPE Asia and Pacific Oil and Gas Conference (1999), Jakarta, Indonesia.

[10] Morillon, A., Vidalie, J.F., Hamzah, S., Suripno, S., Hadinoto, E.K., *Drilling and waste management*, SPE 73931, Proceedings of the SPE International Conference on Health, Safety and the Environment in Oil and Gas Exploration and Production (2002).

[11] Smith, M., Manning, A., Lang, M. *Research on the re-use of drill cuttings onshore.* 1999. Available at http://www.bmtcordah.com/downoads/Re-use%20of%20Drill%20Cuttings%20Onshore.pdf.

[12] EMBRAPA (1997). Manual de Métodos de Análise de Solos (segunda edição), Rio de Janeiro, 1997, p. 212.

[13] EMBRAPA, Sistema Brasileiro de Classificação de Solos, 2009, Brasília, p. 412.

[14] M-CIENTEC, Argilas: Determinação da Contração Linear de Secagem. Método de Ensaio C-021, São Paulo, 1995, p. 3.

[15] M-CIENTEC, Materiais Cerâmicos: Determinação da Contratação Linear de Queima. Método de Ensaio C-026, São Paulo, 1995, p. 6.

[16] American Society for Testing and Materials, Standard Test Method of Soils for Engineering Purposes, Annual Book of ASTM Standards, 04.08, Philadelphia, 1995, p. 1806.

[17] Moore, C.A., Suggested Method for Application of X-Ray Diffraction of Clay Structural Analysis to the Understanding of the Engineering Behavior of Soils. Special Procedure for Testing Soil and Rock for Engineering Purposes. ASTM STP 479, Philadelphia, 1970, 291-300.

[18] United States Environmental Agency. Hazardous Waste Test Methods / SW 846. Toxicity Characteristic Leaching Procedure – Method 1311, Washington, D.C., 1992, p. 35.

[19] American Society for Testing and Materials, Standard Test Methods for Sampling and Testing Brick and Structural Clay Tile – C67, Philadelphia, 2014, 1-16.

[20] Dondi, M., Caracterização tecnológica dos materiais argiosos: métodos experimentais e interpretação dos dados. Cerâmica Industrial, 2006, 11, 36-40.

[21] Balaton, V.T., Ferrer, L.M., Incorporação de Resíduos Sólidos Galvânicos em Massas de Cerâmica Vermelha. Cerâmica Industrial, 2002, 7, 42-45.

[22] Modesto, C., Bristot, V., Menegali, G., de Brida, M., Mazzucco, M., Mazon, A., Borba, G., Virtuoso, J., Gastaldon, M., de Oliveira, A.P.N., Obtenção e caracterização de materiais cerâmicos de resíduos sólidos industriais. Cerâmica Industrial, 2003, 8, 14-18.

[23] Silva, J.R.R., Portella, K.F., Caracterização físico-química de massas cerâmicas e suas influências nas propriedades finais de revestimentos cerâmicos. Cerâmica Industrial, 2005, 10, 12-18.

[24] Alves, M.R.F.V., Holanda, F.S.R., Reciclagem da borra oleosa através da incorporação em blocos cerâmicos de vedação. Cerâmica Industrial, 2005, 10, 17-20.

REUSE OF WASTEWATER FOR IRRIGATION IN MENA REGION AND KONYA EXPERIENCES

[1]**Mehmet Emin Aydin**, [1]**Senar Aydin**, [1]**Fatma Bedük**, [2]**Müfit Bahadir**

[1]Necmettin Erbakan University, Engineering-Architecture Faculty, Environmental Engineering Department, Koycegiz Campus, Meram, Konya, Turkey (meaydin@konya.edu.tr)

[2]Technical University Braunschweig, Institute of Environmental and Sustainable Chemistry, Hagenring 30, 38106 Braunschweig, Germany

Keywords: Irrigation, soil pollution, water scarcity, wastewater treatment, water reuse

Abstract

Arid and semiarid regions receive very little precipitation; therefore, water sources are limited. However, water demand is very high because of high temperature and population. Over 80% of water demand is for agricultural use in Asia and Africa. Due to the urbanization and industrialization of rural areas, the demand for water resources has grown and water becomes increasingly polluted. Wastewater reuse appears to be an available source for irrigation purposes and being practiced in several parts of the world. In this work, general information about water supply, wastewater treatment, effluent discharge and reuse applications in MENA Region as well as in Turkey are exemplified. An overview is given about the wastewater reuse in particular for irrigation purposes in Turkey and the Arab Countries in the Middle East. Annual Mean precipitation amount in Konya is about 325 mm. Surface area of Konya is about 63,757 km^2, which is about 8% of Turkey's total surface area. The amount of land suitable for agriculture is 2,754,243 ha in Konya, while about 835,000 ha of this land could be irrigated at present. Therefore, irrigation water demand is very high. Water usage, wastewater treatment and reuse practices, and experiences of Konya are presented. Some results of a scientific project carried out on pollution caused by long term irrigation with wastewater in the agricultural area of Konya are presented, as well.

1 Introduction

Water scarcity is one of the main challenges in the Middle East and North African Countries (MENA) , where agriculture and domestic needs are the two main water consuming sectors. Recent climate change studies address less precipitation and higher temperature in MENA region. Excessive water losses from drinking water distribution networks are common in many countries. In addition, over 70% of the total available water in MENA countries is allocated to agricultural purposes, where water losses due to inadequate irrigation techniques are also significant. Excessive abstraction of groundwater causes lowering the groundwater table, resulting in saline water intrusion in coastal regions and also reducing surface areas of lakes.

Climate change and population growth will make the water allocation under the water demanding sectors more difficult in the near future [1].

Water scarcity in MENA region has forced countries to reuse reclaimed wastewater for agriculture, industry, recreation, and recharging aquifers. Reclaimed water reuse options are important for both reducing freshwater demand as well as protecting water bodies from wastewater discharge. By this way, it is possible to reserve fresh water resources for domestic uses. Besides, usage of chemical fertilizers may be reduced by using nutrient ingredients N and P of reclaimed wastewater. Using filtration capacity of soil is a good treatment alternative that can be used for reclaimed wastewaters after secondary treatment. Wastewater reuse applications can be applied in a large scale to agricultural irrigation, landscape irrigation, industrial recycling and reuse, groundwater recharge, recreational/environmental uses, non-potable urban uses, and potable reuse. Off all these options the agricultural irrigation is practiced most often [2].

Aside from above mentioned advantages of wastewater reuse for irrigation, there are also potential adverse effects on soil and plants. Electrical conductivity (EC), total dissolved solids (TDS), sodium adsorption ratio (SAR), sodium (Na), chloride (Cl), boron (B), and heavy metals are regulated parameters for the application of wastewater irrigation [3]. Persistent Organic Pollutants (POPs) are another main concern of wastewater irrigation. The irrigation derived POPs in wastewater result in depositions of these pollutants in soil. Hence, this affects the flora and soil fauna, consequently, the human health through food chain. Plant uptake of these contaminants differs according to the type of the plants [4]. Biosolids and sewage sludge are other sources of POPs accumulation in soil [5].

In many MENA countries wastewater is reused in agriculture, e.g., 8,000 ha area is irrigated by reclaimed wastewater after secondary treatment in Gaziantep, Turkey; 500 ha agricultural area is irrigated by effluents of anaerobic and facultative ponds in Al Samra, Jordan; 600 ha area is irrigated by reclaimed wastewater after secondary treatment in La Soukra Irrigation Area, Tunisia. While these examples stand for restricted irrigation of only some agricultural species, effluents of tertiary treatment are being used for unrestricted irrigation of 16,000 ha area in Dan Region, Israel.

Even there is a water stress in some basins in Turkey; the reuse of treated wastewater is not widely acceptable in the public. There is a resistance towards using treated wastewater because of some cultural and traditional reasons. However, nutrient ingredients of waste-water motivate the farmers to go for the use of wastewater for irrigation. In some cases, there is no better alternative. Reuse of treated wastewater is permitted for both non-processed and processed food in Turkish regulations. While secondary treatment, filtration and disinfection are necessary processes for reuse of wastewater for non-processed foods consumed directly and for recreational areas, secondary treatment and disinfection are necessary processes for reuse of wastewater for processed food, non-public green areas such as grass production and grassland.

High number of WWTPs was constructed in Turkey in the previous decade. By the year 2012, 72% of municipal population in Turkey took up wastewater treatment services. There were 595 WWTPs in the year 2014 that treated a total amount of 3.5 billion m^3 (BCM) wastewater. 41.6% advanced, 33.2% biologic; 25% physical, and 0.2% natural treatment methods were used [6]. According to FAO, 0.18% of soils are irrigated with treated municipal wastewaters in Turkey. It is reported that 5.75% of Jordan soils, 1.18% of Saudi Arabian soils, and 17.78% of Israeli soils are irrigated with treated municipal wastewaters. 240,000 ha area in Iran, 665 ha area in Jordan, 2,850 ha area in Saudi Arabia, 40,000 ha in Syria, and 9,160 ha area in Turkey are irrigated with untreated municipal wastewater [7].

In this study, it is aimed to give some results of a research project conducted in Konya, Turkey. In the context of the project, effects of long term irrigation on Konya soils and contamination of crops cultivated in this area wre determined. At the same time, a similar research was performed in Braunschweig, Germany. Aydin et al. [8] published the data about heavy metals accumulation in Konya soil and crops grown on these fields. In this study, accumulation of persistent organic pollutants (POPs), such as polychlorinated biphenyls (PCBs) and polycyclic aromatic hydrocarbons (PAHs), on Konya soils irrigated with wastewater over 40 years are given.

2 Material and Methods

2.1 Study Area and Sampling

Konya city is located in the center of Turkey. Drought is perhaps the most pressing climate concern that faces Konya city. Persistent dry conditions have been evidenced for the past two decades. Konya basin is the largest gross producer of wheat in Turkey. There are flat plains suitable for agriculture with the lack of fresh water. Domestic and industrial wastewaters of the city had been conveyed to the main drainage channel towards the Salt Lake without any treatment until 2010. Since then, wastewaters of the city have been treated in a conventional Wastewater Treatment Plant (WWTP). Konya's WWTP was designed for organic carbon and partial nitrogen (N) removal, including activated sludge basins working by the Bardenpho process.

During the arid seasons, the wastewater in the main drainage channel has been used for land irrigation by farmers without insufficient control. In this study, 27 wastewater irrigated agricultural soil (called WWI soil) samples were taken from near channel in line of 1st, 2nd and 3rd pump stations in a distance of 100 and 500 m from the channel in the depths of 0-25 cm, 25-50 cm and 50-75 cm, in the period of 2011-2012. Three control samples (called Control soil) were also taken from the agricultural soil irrigated solely with well water in the region. 0.5 kg of soil was taken from each sampling point by using screw type steel soil probe and the samples were kept in 4 °C in freezer until analyses.

2.2 Sample preparation, extraction and clean up

Soil samples were air dried, cleaned from stones and passed through a 2 mm sized sieve before analyses. 10 g of soil samples were extracted by 150 m_ n-hexane/acetone (1/1, v/v) solvent mixture for 16 hours, by using soxhlet extraction. Extracts were concentrated to 2 mL by rotary evaporator and gentle nitrogen stream. Column chromatography technic was used for cleanup step by using 10 g 5% deactivated silica gel (US EPA Method 3630C). For the elution of PAHs and PCBs, 70 mL n-hexane and 60 mL n-hexane/ethyl acetate (1/1, v/v) were used, respectively. Extracts were concentrated to 1 mL and subjected to GC-MS analyses.

2.3 GC-MS Analyses

Determination of polychlorinated biphenys (PCB 28, 52, 101, 138, 153, 180) and polycyclic aromatic hydrocarbons (16 PAHs after USEPA, naphthalene, acenaphthalene, acenaphthene, fluorene, phenanthrene, anthracene, fluoranthene, pyrene, benzo[a]anthracene, chrysene, benzo[b]fluoranthene, benzo[k]fluoranthene, benzo[a]pyrene, indeno[1,2,3-cd]pyrene, dibenzo[a,h]anthracene, benzo[g,h,i]perylene) were carried out by gas chromatograph (GC, Agilent 6890N, Agilent Technologies, Palo Alto, CA, USA) equipped with mass-selective detector (MS, Agilent 5973, Agilent Technologies, Foster City, CA, USA). HP-5MS capillary column (30 m length, 0.25 mm i.d. and 0.25 μm film thickness) was used in the GC-MS system. Helium (purity 99.999%) was used as carrier gas at constant flow-rate of 1.9 mL/min. Optimum GC-MS conditions were determined by injection of 1 ng/μL 7 PCB and 16 PAH standards.

2.4 Quality Control

The calibration techniques for PCBs and PAHs were the external standard multipoint calibration using at least five concentration levels. For PCBs, linear regression (R^2 > 0,999) was obtained in the range of 0,001-1 ng/μL. The limits of detection (LODs), based on the signal-to-noise ratio (S/N) were in the range of 0,01 pg/μL (PCB 153) – 0,02 pg/μL (PCB 28, PCB 52, PCB 101, PCB 118, PCB 138, PCB 180). The limits of quantification (LOQs) ranged between 0,04 pg/μL (PCB 153) – 0,08 pg/μL (PCB 28, PCB 52, PCB 101, PCB 118, PCB 138, PCB 180). For the spiking concentration of 0.1 ng/μL, recovery rates after extraction and concentration procedures were in the range of 96-108%. For PAHs, linear results (R^2 > 0,999) were obtained in the range of 0.001-10 ng/μL. LODs were in the range of 0.02 pg/μL (naphthalene) – 2.30 pg/μL (chrysene), and LOQs were in the range of 0.07 pg/μL (naphthalene) – 7.46 pg/μL (chrysene). For the spiking concentration of 0.1 ng/μL, recovery rates after extraction and concentration procedures were in the range of 75-95%.

3 Results and Discussion

3.1. Levels of PCBs pollution in soil samples

PCBs pollution of soil samples taken from in line 1st, 2nd and 3rd pump along the main drainage channel and control soil irrigated with well water are given in Figure 1. Results show that control soil samples contain higher concentration of PCBs. This result indicates PCBs pollution sources other than wastewater irrigation. Pesticides, fertilizers and atmospheric transport are probable sources of PCBs in control soils. Since there is a higher need for fertilizers in well water irrigated soils, higher amounts of PCBs may be introduced with fertilizers. PCB 28 and PCB 52 were in lower concentrations when compared with other analyzed PCBs. The highest concentration was determined for PCB 153 (21 µg/kg). Lower concentrations of PCB 28 and PCB 52 can be explained with lower chlorinated structures and low molecular weights. Higher chlorination results in a lower solubility in water and thus accumulation in soil or sediment.

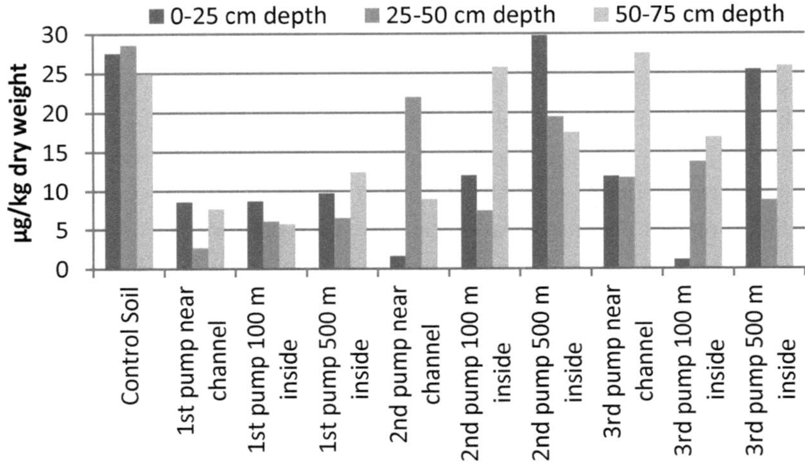

Figure 1: Total PCBs in soil samples

In the literature, PCBs contamination of soils are reported which are very far from the main emission sources. Meijer et al. [9] determined PCBs pollution of non-agricultural soils in Antalya which is in the same range like at the City to Konya. 86% of PCBs worldwide were produced in an area that are in 30° - 60° N-latitude, which Turkey take part in. Different from results of this study, Wang et al. [10] determined PCBs pollution as a result of wastewater irrigation. Total PCBs were determined in the range of 256-2,140 pg/g dry weight (dw) in surface soils, While long range atmospheric transport is the main reason for PCBs pollution, industrial discharges are also sources of PCBs. Konya's municipal wastewater includes nearly 7% industrial wastewater that can be a source for the pollution of soil samples.

3.2. Levels of PAHs pollution in soil samples

PAHs pollution of soil samples taken from in line 1st, 2nd and 3rd pump along the main drainage channel and control soil irrigated with well water are given in Figure 2. Results show that PAHs pollution levels of the whole area are similar. It is obvious that no additional PAHs pollution was determined as a result of wastewater irrigation. The most dominant PAHs were naphthalene, phenathrene, and anthracene, while acenaphthylene, acenapthene, and benzo[a]anthracene were in lowest concentrations.

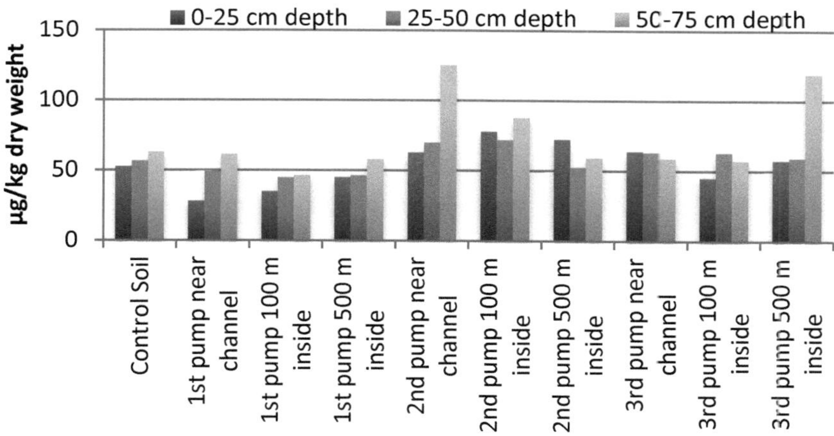

Figure 2: Total PAHs in soil samples

While PCBs were found in the range of 1.12 − 29.7 µg/kg dw, PAHs were determined in a higher range of 27.9 - 125 µg/kg dw. No regular trend, neither for PCBs nor for PAHs, was determined for soil samples taken from different depths, probably because of the soil homogenization through ploughing.

Legal regulations are strengthened in the MENA Region and there are important technical developments for wastewater treatment to meet the regulations. The main challenge in the region is the lack of data to better understand the dimension of the risk caused by wastewater irrigation. There are a few studies reporting effects of POPs contamination of soils. Haddaoi et al. [11] determined occurrence and distribution of 13 PAHs ard 18 PCBs in Tunisian (Nabeul) soil irrigated with treated wastewater. The total PAHs and total PCBs in 0-10 cm top layer of soil were determined in the range of 120 - 365 µg/kg dw and 11.3 - 21.9 µg/kg dw, respectively. Six PCB congeners (28, 52, 101, 138, 153, 180) were determined being dominant.

Long term application of untreated or insufficiently treated wastewater causes not only soil deterioration, but also the contamination of agricultural products. The use of untreated wastewater for edible crops has a potential risk for human health. Khan et al. [12] determined accumulation of PAHs in lettuce grown in the soils irrigated with wastewater, and indicated a potential risk for the consumers of leafy vegetables cultivated in soils irrigated with wastewater.

Microbial quality of wastewater must be one of the main concerns to avoid contamination risk. Vergine et al. [13] carried out a field scale experimental study to determine the fate of *E. coli* in the soil and transfer to groundwater or to the food chain. They found that effects on top soil were strongly dependent on the *E. coli* concentration of wastewater. There were limited persistence in the topsoil, and the trend of the *E. coli* concentrations in the leaching of the soil columns followed a log-linear model suggesting bacterial decay.

Excessive use of wastewater may result in the pollution of underground and surface water bodies. Hidayat et al. [14] determined the impact of wastewater on soil, vegetables, and underground water of Peshawer, Pakistan. A positive correlation was found between heavy metal concentrations, the soil carbon content and the cation exchange capacity of soil. The vertical movement of heavy metals was evidenced, adressing as a risk for underground water pollution.

4 Conclusions

Wastewater irrigation option will probably be more common in the near future as a result of increasing water stress in MENA region. There are numerous advantages of wastewater irrigation, such as soil fertility and crop productivity. However, there are potential adverse effects on physicochemical properties of soil. The quality of treated wastewater, duration of application, method of irrigation and soil physicochemical properties are important factors for safe application of wastewater on soil. It is necessary to monitor soil and plant contamination. Some parameters of a research study conducted to determine PCBs and PAHs contamination of soil samples from Konya, Turkey are given in this study. Even all of soil samples were polluted with PCBs and PAHs, it was not possible to correlate them with long term wastewater irrigation, since the control soils were polluted, as well.

For the safe reuse of reclaimed wastewater for irrigation purposes, it is necessary to remove pathogens and to achieve good treatment results in order to avoid soil salination and food contamination. Minimization of wastewater contact with edible parts of agricultural product is also another issue subject to preserve the products. As a cautious approach, restricted irrigation for limited types of products may be applied. It is essential to consider the soil structure, tolerance of plants towards the respective pollutants, and the type of irrigation methods applied in the decision making process.

5 Acknowledgements

The authors thank to DAAD and Exceed Swindon Project for granting their participation at the *"International Workshop on Wastewater treatment and reuse for Metropolitan Regions and Small Cities in Developing"* held in Recife, Pernambuco, Brazil, on September 13-17, 2016.

6 References

[1]   Al-Saidi, M., Birnbaum D., Buriti R., Diek E., Hasselbring C., Jimenez A., Woinowski D., Water Resources Vulnerability Assessment of MENA Countries Considering Energy and Virtual Water Interactions, Procedia Eng. 2016, 145, 900 – 907.

[2]   Bahadir, M., Aydin, M.E., Aydin, S., Beduk, F., Batarseh, M.. Wastewater Reuse in Middle East Countries – A Review of Prospects and Challenges, Fresenius Env. Bull. 2016, 25(5), 1284-1304.

[3]   NTR, 2010, Turkish Notification of Technical Rules for Wastewater Treatment Plants, Official Newspaper No: 27527, 2010.

[4]   Nasir, F.A., Batarseh, M.I., Agricultural Reuse of Reclaimed Water and Uptake of Organic Compounds: Pilot Study at Mutah University Wastewater Treatment Plant, Jordan, Chemosphere 2008, 72, 1203-1214.

[5]   Armitage, J.M., Hanson, M., Axelman, J., Cousins, I.T., Levels and vertical distribution of PCBs in agricultural and natural soils from Sweden. Sci. Total Environ. 2006, 371, 344–352.

[6]   TUIK, 2015, Turkish Statistikal Agency, http://www.tuik.gov.tr/PreHaberBultenleri.do?id=18778 (Accessed on 13.12.2016)

[7]   Aquastat data, http://www.fao.org/nr/water/aquastat/data (Accessed 13.12.2016)

[8]   Aydin, M.E., Aydin, S., Beduk, F., Tor, A., Tekinay, A., Kolb, M., Bahadir, M., Effects of long term irrigation with untreated municipal wastewater on soil properties and crop quality, Environ. Sci. Pollut. Res. 2015, 22, 19203-19212.

[9]   Meijer, S. N., Ockenden, W. A., Sweetman, A., Breivik, K., Grimalt, J. O., ve Jones, K. C., Global distribution and budget of PCBs and HCB in background surface soils: implications for sources and environmental processes, Environ. Sci. Technol. 2003, 37(4), 667-672.

[10] Wang, T., Wang, Y., Fu, J., Wang, P., Li, Y., Zhang, Q., Jian, G., Characteristic accumulation and soil penetration of polychlorinated biphenyls and polybrominated diphenyl ethers in wastewater irrigated farmlands, Chemosphere 2010, 81, 1045–1051.

[11] Haddaoui I., Mahjoub, O., Mahjoub, B., Boujelben A., Bella G.D., Occurrence and distribution of PAHs, PCBs, and chlorinated pesticides in Tunisian soil irrigated with treated wastewater, Chemosphere 2016, 146, 195-205.

[12] Khan, S., Aijun L., Zhang, S., Hu, Q., Zhu Y.G., Accumulation of polycyclic aromatic hydrocarbons and heavy metals in lettuce grown in the soils contaminated with long-term wastewater irrigation, J. Hazard. Mater. 2008, 152, 506–515.

[13] Vergine, P., Saliba, R., Salerno, C., Laera, G., Berardi, G., Pollice, A., Fate of the fecal indicator *Escherichia coli* in irrigation with partially treated wastewater, Water Res. 2015, 85, 66-73.

[14] Hidayat, U., Ikhtiar, K., Ihsan, U., Impact of sewage contaminated water on soil, vegetables, and underground water of periurban Peshawer, Pakistan, Environ. Monit. Assess. 2012, 184(10), 6411-6421.

GREYWATER REUSE IN JORDAN – COMMUNITY–BASED EXPERIENCES

Ahmad Oqla Alulayyan, Heba A. Ababneh

Mercy Corps, 16, Samirra' St., Amman, Jordan (alulayyan.ahmad@gmail.com)

Keywords: Agriculture, Greywater, Jordan, Reuse, Treatment.

Abstract

Recently, the crisis of water shortage in Jordan has been exacerbated by the climate change, high population growth, frequent influxes of refugees, increased rural to urban migration, and modernization with higher standards of living. Consequently, Jordan is facing a future of very limited water share among the lowest in the world on a per capita basis. This paper describes the possibilities and opportunities of reusing the greywater as a nontraditional water resource for agricultural uses through a pilot project in Mafraq, Jordan. 10 Greywater treatment units were installed in 10 houses. These treatment units were designed according to the quantity and quality of the greywater. The quality of treated greywater varied from sample to sample. The pH varied from 7.41 to 8.93, BOD_5 from 258 to 474 mg/L, TSS from 10 to 256 mg/L, TP from 2.90 to 10.4 mg/L, NH_4 from 49.7 to 414 mg/L, *E. coli* from $3x10^3$ to $2.8x10^6$ MPN/100 mL, and fecal *Enterococcus* from <2 to $2.4x10^2$ MPN/100 mL in the tested samples. In this paper, a case study has been conducted and reported in order to assess the economic and social impacts of reusing greywater. The study revealed that greywater reuse is feasible under specific conditions.

1 Introduction

Recently, the crisis of water shortage in Jordan has been exacerbated by the climate change, high population growth, frequent influxes of refugees, increased rural to urban migration, and modernization with higher standards of living. Consequently, Jordan is facing a future of very limited water share among the lowest in the world on a per capita basis. The significant increase in population (2.6%) has led to a sharp decrease in per capita water availability in Jordan, which dropped from 3,600 m^3 in 1946 to 120 m^3 in 2015 [1,2], and the per capita daily consumption has dropped in the host community areas from over 88 L to 64.5 L [3] . Besides the rapid population growth, the impacts of climate change and the influx of around 3 Million refugees from the neighboring countries will further intense the problem. Temperature will increase and the total annual precipitation is likely to decrease, however, with a fair share of uncertainty. It has been reported that 70% of water resources is used for agricultural purposes, which is essential to achieve the food security regardless the immediate economic revenue of this activity. Finding new and sustainable water resources, wastewater reuse, for instance, emerges as a technical solution to secure water for agricultural production. According to Food and Agriculture Organization [4], the predicted impacts of climate change has increased the crop water requirements, the competition between weed and crops, the spread of pests and nematodes, and the salinization of soils. The combination of increased temperature and decreased rainfall would be expected that result in reducing yield of agricultural crops. Implementing the non-parametric test of Mann–Kendall on historical climatic records of Jordan showed decreasing trends (8–20%) in precipitation and increasing trends for temperature (1.0–2.8 °C) in the northern and western parts of Jordan [5]. A recent article in the American Journal of Environmental Sciences comes to similar results [6].

Jordan also has witnessed a significant population increase due to the influx of refugees from the neighboring countries. The recent statistics revealed that Jordan hosts 1.265 million Syrian refugees, which comprise 46% of the total population [7]. This consequently affects the demand on the country's very limited resources including water, and affected the economic situation in the country as the per capita GDP growth has fallen by 64% [3]. The increasing demand on freshwater puts more pressure on the limited water resources. The most extensive consuming sector for water resources is agriculture. Pressure on water resources is particularly acute in arid regions with large population where water use is high relative to water availability [8].

Wastewater reclaiming emerges as a technical solution to mitigate or to reduce the consequences of water scarcity. Recycling of wastewater is one of the main options when looking for new water resources in a water scarce region [9], and is considered a significant step towards sustainable development and sound environment management. Greywater is urban wastewater that includes water from baths, showers, hand basins, washing machines, dishwashers, and kitchen sinks, but excludes streams from toilets [,10,11,12,13], and contains 50–80% of the total household wastewater [14,15]. On household level, 65-100% of household wastewater is greywater, i.e. wastewater without urine or faeces content [16] (Table 1). This shows that greywater treatment and reuse for irrigation of forage crops and olive trees is an applicable, reliable and inexpensive solution to help the rural and suburban households in Jordan to keep their income generating activities.

Table 1: Measured values of nutrients in greywater [18]

Nutrients [mg/L]	Laundry	Bathroom	Kitchen sink
Ammonia (NH_3-N)	< 0.1- 3.47	<0.1- 25	0.2- 23.0
Nitrate and nitrite as N	0.10- 0.31	<0.05- 0.20	-
Nitrate (NO^3-N)	0.4- 0.6	0- 4.9	-
Phosphorus as PO_4	4.0- 15	4- 35	0.4- 4.7
Nitrogen as total	1.0- 40	4.6- 20	15.4- 42.8
Tot- N	6- 21	0.6- 7.3	13- 60
Tot- P	0.062- 57	0.11- 2.2	3.1- 10

Greywater is also considered a good source for plant nutrients and thus reduces the costs of fertilization. According to [17], greywater has a typical N/P ratio of 2, thus far below the N/P ratio of around 10 which would be optimal for nutrient uptake by plants. This is very important if greywater is reused for irrigation. Nitrogen then represents the limiting factor, leading to a sub-optimal phosphorus uptake unless the plants can get nitrogen from other sources.

Due to the low levels of contaminating pathogens and nitrogen, reuse and recycling of greywater is receiving more and more attention [19]. Numerous studies have been conducted on the treatment of greywater with different technologies, which vary in both complexity and performance. However, specific guidelines for greywater reuse are not available or not sufficient, and studies on the evaluation of the appropriate technologies for grey water reuse/ recycling are scarce. However, the advantages of wastewater treatment and reusing system may be combined with a number of risk factors. Some risk factors are short term and vary in severity depending on the potential for human, animal or environmental contact (e.g., microbial pathogens), while others have longer term impacts which increase with continued use of recycled water (e.g., salinization effects on soil) [20].

Most greywater is generated by bathing, cooking and laundry, and has low concentrations of nitrogen and variable concentrations of organics, salts, fat and oils, and surfactants [21]. But major risks affecting the reuse of greywater is that most household chemicals end up in the sinks and laundry, which can have negative or positive effects if the water is used for irrigation. Therefore, treatment is necessary to reduce the biochemical oxygen demand (BOD_5), to prevent the anaerobic situation, which may lead to environmental problems such as bad odor and toxicity to animals and plants. Removal of surfactants is especially important, as they can accumulate in soil and even at low concentrations of ~20 mg/kg virtually eliminate the capillary rise of water in sandy soil, giving it hydrophobic properties [22]. The concentration of pathogens in greywater is on average 104 times lower than in mixed wastewater [23]. This greatly simplifies the treatment needed before the greywater can be safely used for crop irrigation, floor cleaning, dust prevention, etc. [24]. However, the pathogen content is still high enough in the water to cause health risks if used untreated [25].

In this paper, a case study has been conducted and reported in order to assess the economic and social impacts of reusing greywater. The study revealed that greywater reuse is feasible under specific conditions.

2 Material and Methods

Mercy Corps-Jordan through its Community-Based Initiatives for Water Demand Management project has provided a grant to install 10 greywater gravel filters in the North-East Badia district in Jordan in cooperation with local Community-Based Organization (CBO). The CBO acted as a community mobilizer to identify the environmental challenges that face the target population and to suggest the most suitable solutions. The target area is recognized as an arid area that receives less than 200 mm precipitation per year. The CBO also assisted in identifying the beneficiaries from this activity. Therefore, this activity was selected based on a participatory way to assure that it meets real needs, enhance the ownership sense, and ultimately achieve the intervention sustainability.

Household selection criteria
- *Household size* to secure constant and reliable flow of the greywater. The average family size in the benefited households was 7.4.
- The *space availability* to construct the greywater treatment system and to reuse the treated greywater (for irrigation in this case).
- *Household's capacity* to cover the cost of the system, a revolving loan mechanism was established to enable the beneficiaries to cover the cost of the system.

Greywater treatment unit design
The greywater treatment unit consisted of 100 L plastic barrel followed by a gravel filter and a collection tank equipped with a pump (Figure 1).

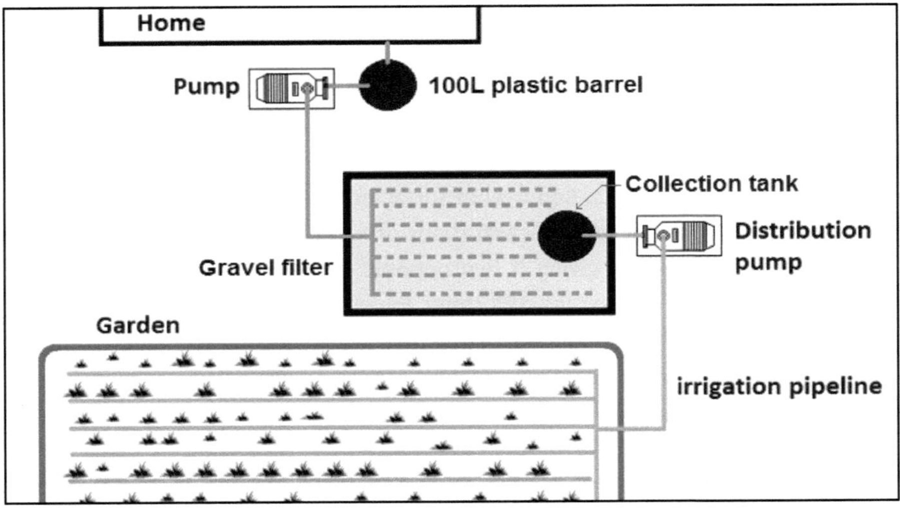

Figure 1 Greywater Treatment System

The greywater conveyed to the 100 L plastic barrel by gravity and then pressurized and pumped to the gravel filter. The filter size was determined based on the organic matter content of the effluent. The maximum total biological loading for the effluent was measured at 172 g BOD_5/day. According to literature, the average organic load per 1 m^2 is 40 BOD_5/day [26].

This study was carried out on one of the constructed greywater treatment systems. The flow rate was measured on daily basis with an average flow of 490 L/day. Following filtration, the greywater was collected in a collection tank equipped with a pump (Figure 1). The treated greywater was used for drip irrigation of the household's backyard. The untreated greywater samples were taken from barrel that received water during 24 h, while treated greywater samples were collected from the collection tank after treatment. Collected samples were analyzed for pH, total suspended solids (TSS), biological oxygen demand (BOD_5), Ammonium (NH_4), and total Phosphorus (TP). Analyses of grey water were carried out following the standard methods for the examination of water and wastewater (Table 2).

3 Results and Discussion

The greywater generation rate in the selected household was measured at 70 L/Capita.day in average. This rate is high compared with rates reported from Amman, 59 L/Capita.day [27]. The low BOD_5 and TSS values in the untreated greywater are mainly attributed to this relatively high water consumption rate. The average removal efficiency for BOD5, TSS, NH4 and TP using the gravel filter was satisfactory compared with those reported by the popular examined greywater treatment systems (Table 3). Most importantly, the values for the determined parameter were acceptable compared with the Jordanian standards for reclaimed wastewater reuse for tree crops' irrigation (effluent quality) except for *E. coli* (Table 4).

Table 2: Filter removal efficiencies

Sample	BOD_5 removal (%)	TSS removal (%)	TP removal (%)	NH_4 removal (%)
S1	74.53	91.41	57%	84.72
S2	89.02	84.91	51%	63.78
S3	96.35	83.17	55%	78.23
S4	96.55	89.05	35%	87.45
S5	85.04	-40.0	39%	98.07
S6	95.35	77.08	53%	94.08
S7	91.1	77.5	11%	93.07
S8	86.69	69.05	39%	79.13
S9	98.3	94.23	73%	94.49
S10	98.49	94.51	71%	93.77
Average	**91**	**72%**	**48%**	**87%**

Table 3: Jordanian standards for reclaimed wastewater reuse in Agriculture

Parameter	Maximum allowable concentration
BOD_5 (mg/L)	200
TSS (mg/L)	150
pH	$6 - 9$
E. coli (MPN/100 mL)	1000
TP (mg/L)	30

Table 4: Untreated and treated greywater characteristics (untreated vs. treated greywater)

Sample	pH		BOD$_5$ mg/L		TSS mg/L		TP mg/L		NH$_4$ mg/L		E. coli MPN/100 mL	
	Untreat.	Treated	Untreat.	Treated	Untreat.	Treated	Untreat.	Treated	Untreat.	Treated	Untreat.	Treated
1	8.93	7.61	371	94.5	256	22	10.4	4.5	83.1	12.7	3×10^5	8.0×10^5
2	8.08	7.76	458	50.3	159	24	5.33	2.6	49.7	18	$8.0\mathrm{X}10^5$	7.0×10^4
3	7.96	7.88	258	9.42	101	17	4.39	1.96	58.8	12.8	$2.8\mathrm{X}10^6$	9.0×10^4
4	7.82	8.1	371	12.8	137	15	2.95	1.91	76.8	9.64	$3.0\mathrm{X}10^5$	$1.3\mathrm{X}10^4$
5	7.95	7.71	274	41	10	14	6.13	3.73	414	8	$3.0\mathrm{X}10^3$	$9.0\mathrm{X}10^5$
6	7.96	8.08	318	14.8	48	11	7.96	3.73	76	<4.5	$7.0\mathrm{X}10^5$	$2.4\mathrm{X}10^5$
7	7.41	7.48	345	30.7	80	18	5.44	4.82	64.9	<4.5	$5.0\mathrm{X}10^6$	$9.0\mathrm{X}10^5$
8	7.57	7.98	320	42.6	84	26	5.68	3.49	78.6	16.4	$3.4\mathrm{X}10^5$	$5.0\mathrm{X}10^5$
9	7.95	7.87	420	7.15	156	9	6.08	1.62	81.7	<4.5	$2.4\mathrm{X}10^6$	$5.0\mathrm{X}10^2$
10	7.66	7.9	474	7.18	164	9	8.57	2.51	72.2	<4.5	$2.4\mathrm{X}10^3$	$8.0\mathrm{X}10^2$

The level of *E. coli* in the untreated greywater varied widely (from 2.4×10^3 to 5.0×10^6), making it difficult to determine the reduction capacity of the filter. A secondary growth of the bacteria could have occurred in the gravel filter or in the post-treatment collection tank. The macro and micro pore size may provide a good environment to the growth of *E. coli*. The World Health Organization did not recommend standards for fecal coliform content in treated greywater used for irrigating cereal crops, industrial crops, fodder crops, pasture and fruit trees. However, In the case of fruit trees, irrigation should be ceased two weeks before the fruit is picked. And sprinkler irrigation should not be used.

4 Conclusions

In arid areas like the study area, the food production depends heavily on supplementary irrigation. Given the fact that the households in this area receive intermittent water supply, they have no enough water for the regular domestic uses, not to mention water for the irrigation purposes. Purchasing water from private sources (springs and private groundwater wells) is common; however, it is unaffordable for many households. Despite the spaces availability, such situations make it hard for the local population to pursue any agricultural activity due water scarcity and its high price.

Greywater treatment and reuse provided a technical solution to increase the water use efficiency and ultimately overcome the socioeconomic impact of water scarcity. It has been reported that adopting this technique reduced the water bills for the households by 27% and increased the olive trees production by 30%. Finding renewable and reliable water source enabled the targeted population to offset the food purchases and generate extra income by selling surplus products. The household-based agricultural practices were mainly performed by women and, therefore, they were the key actors and beneficiaries. This led to women power enhancement through the household-based enabling income generating environment.

It is worth mentioning that the target area is not connected with the municipal sewer network and, therefore, depends on unlined septic tanks. Emptying these septic tanks is carried out on monthly basis and costs around 100 USD monthly. Adopting the greywater treatment systems decreased the wastewater flow to the septic tanks and decreased the cost associated to its emptying significantly.

5 Acknowledgements

Authors would like to thank DAAD and Exceed Swindon Project for providing this opportunity to present this work at the expert workshop 2016 in Recife, Brazil and for the associated financial support they provided. Authors are also grateful to the local community of the North-East Badia District, who participated in this pilot project and contributed to its success.

6 References

[1]   Ministry of Water and Irrigation, Water for life - Jordan's Water Strategy 2008-2022, Amman, Jordan, 2008.

[2]   Ministry of Water and Irrigation, National Water Strategy 2016-2025, Amman, Jordan, 2015

[3]   Ministry of Planning and International Cooperation, Jordan response platform, and United Nations, Jordan Response Plan for the Syria Crisis 2016-2018, 2016,.

[4]   FAO - Food and Agriculture Organization of UN, Impact of Climate Change, Pests and Diseases on Food Security and Poverty Reduction. Special Event Background Document for the 31st Session of the Committee on World Food, 2005.

[5]   MoEnv (Ministry of Environment), Jordan's Second National Communication to the UNFCCC. Ministry of Environment, Amman, Jordan, 2009.

[6]   Hamdi, R., Abu-Allaban, M., Al-Shayeb, A., Jaber, M. and Momani, N., Climate Change in Jordan: A Comprehensive Examination Approach, American Journal of Environmental Sciences 5 (1): 58-68, 2009.

[7]   Department of Statistics, Amman-Jordan, 2016.

[8]   Mygatt, E., World's water resources face mounting pressure. Eco-Economy Indicators. Retrieved from www.earth-policy.org/index.php?/indicators/C57, 2006.

[9]   Blumenthal, U.J., Mara, D.D., Peasey, A., Ruiz-Palacios, G., Scott, R., Guidelines for the microbiological quality of treated wastewater used in agriculture: Recommendations for revising WHO guidelines. Bulletin of the WHO 2000, 78(9), 1104-1116.

[10]   Jefferson, B., Laine, A., Parsons, S., Stephenson, T., Judd, S., Technologies for domestic Wastewater recycling. Urban Water, 1999, 1(4), 285–92.

[11]   Otterpohl, R., Albold, A., Oldenburg, M., Source control in urban sanitation and waste management: ten systems with reuse of resources. Water Sci. Technol. 1999, 39(5), 153–60.

[12]   Eriksson E., Potential and problems related to reuse of water in households. Ph.D. Thesis. Environment and Resources DTU, Technical University of Denmark, 2002, ISBN 87-89220-69

[13]   Ottoson, J., Stenström, T.A., Faecal contamination of Greywater and associated microbial risks. Water Res. 2003, 37(3), 645–55.

[14]   Eriksson, E., Auffarth, K., Eilersen, A-M., Henze, M., Ledin, A., Household chemicals and personal care products as sources for xenobiotic organic compounds in grey wastewater. Water S A 2003, 29(2), 135–46.

[15]   Friedler, E., Hadari, M., Economic feasibility of on-site grey water reuse in multi-storey buildings. Desalination 2006, 190(1-3), 221–34.

[16]   Morel, A., Diene,r S., Greywater management in low and middle income countries. Sandec, Eawag, Switzerland, 2006. ISBN: 978-3-906484-37-2.

[17] Gunther, F., Wastewater treatment by Greywater separation: Outline for a biologically based Greywater purification plant in Sweden. Ecol. Eng., 2000, 139-146.

[18] Ledin, A., Eriksson, E., Henze, M., Aspects of groundwater recharge using grey wastewater. In: Decentralised Sanitation and Reuse, G. Lettinga, (ed.), London, 2001 650.

[19] Li, Z., Gulyas, H., Jahn, M., Gajurel, D.R., Otterpohl, R., Greywater treatment by constructed wetland in combination with TiO_2-based photocatalytic oxidation for suburban and rural areas without sewer system. Water Sci. Technol. 2003, 48(11), 101–106.

[20] Toze, S., Water reuse and health risks – real vs. perceived, Desalination, 2006, 187(1-3), 41-51.

[21] Shafran, A.W., Gross, A., Ronen, Z., Weisbrod, N., Adar, E , Effects of Surfactants Originating from Reuse of Greywater on Capillary Rise in the Soil. Water Sci. Technol. 2008, 52(10-11), 157-166.

[22] Shafran, A.W., Ronena, Z., Weisbroda, N., Adara, E., Gross, A., Potential changes in soil properties following irrigation with surfactant-rich Greywater. Ecol. Eng. 2006, 26, 348–354.

[23] Ottosson, J., 2005. Comparative analysis of pathogen cccurrence in wastewater. PhD Thesis. Royal Institute of Technology, KTH, 2005

[24] WHO, Overview of Greywater management: health considerations 2006.

[25] WHO. WHO guidelines for the safe use of wastewater, excreta and Greywater/ World Health Organization. Geneva: World Health Organization. 2006, 182

[26] S. Sinthana Gorky, Treatment of Greywater using constructed wetlands. International Research Journal of Latest Trends in Engineering and Technology (IRJLTET). Vol. 2(3) 2015

[27] Jefferson, B., Laine, A.L., Stephenson, T., Judd, S.J. Advanced biological unit processes for domestic water recycling. Water Sci. Technol. 2001 43(10) 211-218.

CLIMATE PROTECTION THROUGH SUSTAINABLE WASTE MANAGEMENT

[1]Christiane Pereira, [1]Klaus Fricke, [2]Jens Giersdorf

[1]Leichtweiss Institute, Department of Waste and Resources Management, Technische Universität Braunschweig, Beethovenstrasse 51 a, D-38106 Braunschweig, Germany (klaus.fricke@tu-bs.de, chrdiasp@tu-bs.de)

[2]Deutsche Gesellschaft für Internationale Zusammenarbeit (GIZ) GmbH, SCN Quadra 01, Bloco C, Sala 1.501, Ed. Brasília Trade Center, 70.711-902 Brasília – DF, Brasil (jens.giersdorf@giz.de)

Keywords: Capacity building, climate change, know-how transfer, waste management

Abstract

In 2010, the National Solid Waste Policy (PNRS) of Brazil was issued, which is based on principles concerning sustainable solutions for problems associated with waste management, resources preservation and climate protection. Therefore, the PNRS creates a positive agenda for the encouragement of adaptation of landfills as electric power plants; closing and remediating wild dumps and promoting waste valorization, recycling and social inclusion. All these new proposals should face the Brazilian reality that goes from economic feasibility discussion over the lack of knowledge to implement sustainable solutions. This fact is not due to lower market interest in the subject, but rather due to the pioneering condition, with no large-scale examples that give opportunity for experiences' exchange. Furthermore, the importance of waste management for greenhouse gas (GHG) emissions and the availability of climate protection measures are hardly known. Based on this need, the proposed international climate initiative (IKI) project aroused a lot of attention in the market, resulting in the formation of strategic and multidisciplinary partnerships for the purpose of democratization of data as well as the development of joint projects that minimized errors, and optimizing the arrangements in favor of consistent projects. The results are still modest but encourage an innovative approach based on integration of consolidated market with new and great opportunities in Brazil.

1 Introduction

Waste management has changed significantly in the past years, becoming an icon of sustainable development, contributing to environmental protection, and guaranteeing climate protection and preservation of natural resources. In this context, Brazil issued a National Solid Waste Policy (PNRS) in 2010, promoting the introduction of selective waste collection and reverse logistics, recycling of household waste, composting of organic waste, and generating renewable energy through biomass and biogas before final disposal throughout the whole country. The implementation of these provisions introduces a new system for waste manage-

ment to the country, particularly in relation to the introduction of technologies for the recovery of waste, offering a series of new activities to be implemented in short and medium term
as established by law. Although the responsibility of the public cleaning services is municipal,
the national regulation strengthens the society and private sector in the requirement of a new
way of dealing with the matter [1]. This poses a considerable challenge due to the limited
expertise available for the necessary technologies, and to streamline them into the Brazilian
market, which in turn results in hesitations of decision-making at all public levels (federal,
state and municipalities) as well as other relevant stakeholders like funding and environmental
licensing agencies.

Although environmental protection found its way into regulations a long time ago, the topic
"climate protection" was not considered by the Brazilian political systems so far. Because of
this fact, there is no specific calculation basis for GHG emissions in Brazil, and the influence of
different sectors on the generation matrix of GHG emissions is not known correctly. The
importance of the waste management sector, and the potential and availability of climate
protection measures can, therefore, hardly be quantified. Hence, the IKI project demonstrates
the potential of waste management for the mitigation of GHG emissions and describes a
German-Brazilian cooperation project that promotes climate protection through introducing
measures of sustainable waste management.

2 Climate change and waste management

Waste management has a high potential for the mitigation and reduction of GHG emissions. In
the period 1990-2012, an amount of 321 million tons CO_2eq was reduced in Germany (from
1,247 to 926 million tons CO2eq), of which waste management measures contributed around
55 million tons CO_2eq [2, 3].

Recent studies showed that the total mitigation potential of GHG emissions released by Brazilian landfills sums up to approximately 42 million tons CO_2eq. Further reduction benefits in the
order of 15 million tons CO_2eq are possible through recycling measures and the related energy
savings. Further savings or even credits up to 5.5 million tons CO_2eq can be generated through
the direct conversion of waste biomass to energy by incineration processes. These performances can be realized despite of the comparably low grid factor of 268 CO_2/MWh that has a particularly negative effect on the climate-friendly performance of recycling and energy-related
services compared with Germany (see Table 1). Since only the major Municipal Solid Waste
(MSW) fractions were analyzed for this estimation so far, these results should be regarded as
the lower threshold, just like the conservative recycling rates of 50%. Modern sorting technologies like near-infrared spectrometry (NIR) reach recycling rates above 80%.

Currently, the annual increase of MSW in Brazil is 3.5%. Through continuous expansion of
wastewater collection and treatment, the GHG potential will continue to multiply due to the

retention of landfills. Hence, the greatest reduction potential provides the sector landfilling of untreated waste. As landfills with high environmental standards normally only capture a share of 30% of landfill gas, but still produce liquid and gaseous emissions for up to 50 years, sustainable waste management measures should not focus on landfilling only, but rather on recycling, energy recovery, and waste treatment before landfilling as a key to climate protection. In addition to the avoidance of GHG emissions from landfill sites, a considerable amount of CO_2 credits can also be generated through recycling and energy recovery from waste and biomass components.

Table 1: Reduction potential of GHG emissions in Brazil through waste management measures

GHG	Waste/ Sludge amount in Brazil	GHG Emissions from Landfill	GHG saving by Recycling[3]	GHG saving by Energy Recovery (Biomass)	Total Potential
	(million tons)	(million tons CO_2 eq)			
MSW	76.4	42.2	6.29 plastics 3.54 metals 5.45 paper cardboard	5.5	63
Sewage sludge	0.22 dm[1] 2.2 dm[2]	0.8 8.9	?	0.25 2.46	1.1 11.4

[1] today, [2] in 2030, [3] by energy saving

3 Waste Management and Global Challenges

3.1 Waste management

The amount of solid waste produced by the population is not only related to the level of wealth reflected in the economic ability to consume, but also to the values and habits of life, determining the degree of disposition for the realization of consumption. In order to reduce the impacts of improper disposal, every effort should be put into action for not producing waste prior to the concerns about its disposal. The background of waste generation and recovery is based on:

➢ The increment of global gross domestic product (GDP), resulting in higher consumption and greater demand for food and primary resources for e.g. fossil fuels,
➢ The increase of greenhouse gas emissions,
➢ Landfills contribute significantly: they represent 8 to 12% of the entropic emissions,
➢ The increase of secondary resource prices as well as energy prices, and the demand for compost are also motivations for waste recovery, and
➢ Directly connected with Brazil, there is also a dependence of energy matrix on climate change.

In order to change traditional practices one needs to open a multidisciplinary discourse for integrating the multiple market segments for enabling the design of those for the implementation of sustainable management of MSW. The discussions range from technologies like fermentation, composting, recycling, and energy recovery to the supply of information. Furthermore, they advise to introduce sustainable waste management and engineering in a scientific context as well as relevant aspects for implementation of projects like financing, environmental licensing, monitoring, and other aspects of the market.

3.2 IKI project description

IKI is the acronym for *Internationale Klima Initiative* (international climate initiative). Since the release of the waste policy PNRS the Brazilian market realized many partnerships and cooperation with international players like the Technical University of Braunschweig (TUBS), the GIZ, the DAAD and the KfW, among other institution. The partnerships aim to improve the municipal waste management in Brazil through promotion of technical capacity development in workshops, international congresses, consulting activities and training, e.g. the project i-NOPA, funded by the DAAD. This project is entitled *"Capacity development and fundamental research in order to create an analyses methodology for the development of a project for the installation of a Mechanical-Biological-Treatment plant (MBT) with integrated fermentation in Jundiaí, SP"*. This intervention gained great interest and increased the representativeness of waste management in the Brazilian economic chain. Various lines of action were promoted, such as equalization of environmental technologies, integrated and sustainable management systems, social inclusion, viability studies, know-how transfer, and formation of a market for sub-products like compost and recyclables as well as alternative energy sources like biogas and RDF (refused derived fuels). However, there were no profound discussions regarding the nexus of climate protection and waste management.

According to official data from the Ministry of Science and Technology, the waste sector in Brazil produced 2016 an amount of around 32 million tons CO_2eq with a rising trend, while estimations from TU Braunschweig have shown a reduction potential of at least 58 million tons CO_2eq through more sustainable measures in waste management. In order to meet the needs for the integration of waste management and climate protection, the German institutions GIZ and TU Braunschweig elaborated the project *"Environmentally appropriate technologies and development of capacities for the implementation of the National Solid Waste Policy PNRS in Brazil."* This project emerged within the scope of the IKI promoted by the *"Federal Ministry for the Environment, Nature Conservation, Building and Nuclear Safety (BMUB)"* and was already approved by partners such as the Brazilian Ministries of Environment and Cities.

Regarding waste and pollution, the current situation in Brazil is in a critical state. Many municipalities count innumerous contaminated sites and high grades of water pollution because of illegally deposited waste. Until 2014, only a share of 58% of municipalities managed to rehabilitate their wild dumps into sanitary landfills or constructed new landfills without any previous waste treatment. This means that a share of 42% of totally 80 million tons of MSW per year is still disposed of inappropriately. Therefore, the extension of sanitary landfill sites is the main effort aimed at municipal and state levels. In order to reach the aims and to utilize the potentials for climate protection in the Brazilian waste sector aspects like waste treatment before final disposal, increase of recycling rates, adaptation and improvement of financial frameworks for municipalities, and capacity development at ministries, public and private banks and especially municipalities must be addressed.

In the past years, the Brazilian government promoted the construction of new landfills by a subsidy of 100% in economically underdeveloped municipalities and by doing so simultaneous impeded measures with greater waste management goals in other municipalities. In few states, environmental agencies approve new landfill sites only with a corresponding treatment plan. In the past, even simple sorting or composting plants were not operated efficiently and, therefore, closed. Today, because of these facts, many municipalities have hesitations regarding the introduction of new technologies for the treatment of MSW.

With the objective to better support municipalities that are willing to invest into MSW treatment management and technologies, and to mitigate GHG emissions, the Brazilian Ministry of Cities and the Brazilian Ministry of Environment presented a project proposal for the 2016 call of the IKI of the German Ministry for Environment (BMUB). The project was approved by the joint Brazilian-German selection committee in 2015 during a visit of German Chancellor Angela Merkel and will be implemented by the GIZ (Deutsche Gesellschaft für Internationale Zusammenarbeit) and TU Braunschweig from April 2017 to March 2021.

The project aims to improve the framework conditions for the utilization of climate protection potentials of the waste sector. Therefore, climate relevant criteria must be integrated in the regulations of ministries. Additionally, municipality and company employees will be trained as well as decision support for municipalities will be provided (*capacity development*) for a successful introduction of sustainable waste management. During the project course, all significant players like representatives of municipalities, associations, cooperatives as well as public and private companies will be integrated in order to achieve an intensive collaboration. Moreover, throughout the integration of climate-relevant aspects into the guidelines of the National Waste Policy, it will be possible to quantify specific reduction goals for the waste sector. Thus, it is important that a national policy for waste should not only include short-term but also medium and long-term measures that could strengthen Brazil's lead role in climate-friendly waste management in Latin America. From a social and economic point of view, the project provides an impulse for the development of an integrated circular economy within the scope of modern waste treatment technologies and strategies. Moreover, this initiative supports further development of the Brazilian economy for a Green Economy and Green Jobs. The expansion of waste collection systems and the partly automated separation and treatment of municipal waste enable a better quality and larger quantity of different available waste fractions, indirectly involving hereby 500,000 informal waste collectors, who are integrated into the collecting and recycling systems by the municipalities.

Regarding the ecological aspects, it is obvious that recycling and energy recovery of waste enable a significant reduction of fossil energy consumption. By encouraging the introduction of separate collection and stabilization (composting and fermentation), and utilization of the organic fractions as fertilizer and humus generator, the quality of Brazil's nutrient and humus poor soils as well as of groundwater will be improved, negative impacts through insufficiently controlled landfills will be minimized, and land consumption for new landfills will be reduced due to the increase of volume and lifespan of existing landfills.

4 Results and Discussion

4.1 Waste management overview

As a consequence of the distinctive population growth of more than 11.9% and an achieved increase of the GDP per capita of 24.4% during the period of 2003-2014, the associated waste generation increased by a rate of 29.6% to about 80 million tons per year in Brazil. Currently, only an amount of 58% of MSW is correctly disposed of on sanitary landfills, while the recycling rate achieves a percentage of only 3-4%. This situation causes great stress on the environment like large contaminated areas, water contamination, mass movements, a contribution of 8-12% to GHG emissions as well as a big loss of materials and secondary resources.

Table 2: Waste Panorama for Brazil [4]

WASTE PANORAMA 2015	
Waste generation	~ 80 million tons
Generation per inhabitant	0.97 kg/inhab/d
Final disposal in sanitary landfills	58.7%
Jobs - private operators (200 companies)	353,328

Waste collection services are provided in 90.8% of Brazilian cities, and about 69% of the municipalities have already implemented a selective collection of their MSW in order to encourage recycling activities. The national recycling of civil construction waste is conducted in 158 municipalities, whereas the recycling rate decreased by 7% in the past year. The index for composting achieved a rate of only 1.6% in 222 municipalities even though the organic fraction in household waste reaches an average of more than 50%. Recyclable materials like paper/cardboards/Tetra Packs and plastics represent 13%, respectively. Although the political attitude towards waste management has changed in the past years, the Brazilian market shows a slow reaction. For instance, during 2010-2014 only 10 contracts dealing with waste recovery were signed. Furthermore, 47% of municipalities struggle with high costs for the municipal cleaning services (~ 8 billion €/yr or ~ 35 €/inhab*yr), which should be covered by a waste tax. Mostly, this tax covers only 40% of the total costs for the services provided by the municipalities. Table 2 contains further information about waste management in Brazil 2015.

The PNRS not only comprises goals for the generation of energy from waste, but also requirements for shared responsibility of secondary products, life cycle analysis, reverse logistics of waste, and specified targets for the reduction of recyclable materials on landfills, as can be seen in Table 3.

Table 3: Goals of the National Solid Waste Plan [1]

WASTE MANAGEMENT REDUCTION TARGETS	4 YEARS [%]	8 YEARS [%]	12 YEARS [%]
Dry fraction	22 – 37	28 – 42	34 – 50
Wet fraction	19 – 35	28 – 45	38 – 55

Along with the PNRS, new opportunities for the management of waste as well as secondary positive effects have been created for the Brazilian main market players, e.g. market searches for an appropriate infrastructure, technologies, and other effective systems incl. technical-operational aspects. Especially the latter include technologies for implementation and monitoring of future waste treatment plants. Therefore, partnerships are formed among national waste operators and suppliers of foreign technologies as well as with new actors in cement, cellulose and energy industries.

Table 4: Market situation for secondary resources

Secondary Resources	Market situation	Comments
Recyclables	Average price 200 Euro/ton	Based on over 370 cooperatives and large informal sector – small scale plants – 150 tons/m
Compost	30-50 Euro/ton for selective waste Actual production 6 million tons/a. Stable market with growth potential	Actually no limitation for mixed waste but is under preparation a new law that will prohibit mixed waste compost
RDF	Average price 20 Euro/ton	Cement industry being business partner, long term contracts Strategy: 30% substitution of energy source in 5 years
Biogas	Average price 45 Euro/MW	Volatile tendency, actually high potential

Nonetheless, the market for secondary resources in Brazil already exists, as Table 4 illustrates. Based on cooperatives and the big informal sector, the prices for recyclables average a price of 150-200 Euro/ton, while the stable market for compost shows a growing potential with an annual production of 6 million ton/a just as important as the production of Biogas.

To sum up, the tendencies show that the Brazilian market for waste management demands new technologies for treatment plants, which are not available in Brazil up to now and, therefore, have to be imported from Germany, France, Portugal or Spain. During the first years of the PNRS, the demand for MBT with integrated bio-digestion was higher than today, which is now rising for RDF and sorting plants for recyclables. Likewise, the cement industry pushes waste recovery. Another proposal for the improvement of the current market situation is the nationalization of treatment plants resulting in commercial flexibility and better financing possibilities.

Furthermore, the Brazilian market is heated due to the provision of public concessions that come to 2 billion Dollars for cities with a population of more than 800,000 inhabitants. There are Brazilian financing lines with various rates, approx. 6.5-12%/yr. Waste management does not only involve a good infrastructure, but also auxiliary public policy. Therefore, it is crucial that potential technologies are consolidated in order to obtain environmental licenses. A foundation for further MBT technologies has been established and, as a consequence, it is necessary to develop and to strengthen the secondary resources market in Brazil.

During the first 2 years after publishing the legal standards, a series of contracts in public-private-partnership (PPP) modalities were signed contemplating extremely complex technologies like incineration and fermentation plants as well as other technologies like sorting of recyclable materials and extensive systems for composting. However, economic analyses performed and energy prices much more reduced than expected, a change in the profile of acquisition of the market was identified, prioritizing more simplistic technologies, such as the recycling of materials. Currently, Brazil has a new market orientation, where implementing of MBT plants is evidenced, having an emphasis on production of RDF to supply the new demands of the cement industry. In this scenario, a sector based on quest for operational excellence has been developed dictating the new rules for an industry accustomed to practices of low complexity such as collection and landfilling of solid waste. This partnership represents a change of paradigm, not only to introduce a new feature to the secondary waste market, but basically to ensure the professionalism of a market used to be highly profitable in short-term, which are no longer in resonance when there are legal requirements for waste recovery. Even skipping these discussions, there is a participation of peripheral actors, such as public prosecutor, financing banks, environmental agencies, and the community, pressing the market to offer its services in accordance with the regulations of the law.

4.2 Current market situation for secondary resources

The sustainable market, which the management of waste is a part of, has in Brazil a positive outlook in the coming years, mainly because of the potential scale of recovery of waste promoted by the National Policy for Solid Waste. The understanding has been established that the proper disposal of waste is not confined with landfilling of waste in nature, defining practices ranging from the implementation of systems for waste recovery to the landfilling of so-called rejects. The majority of waste is organic that should be composted, and the recyclable portion should be removed through selective collection or MBT plants. The legal framework under federal, state, and municipal competence will fix the percentage of residues that should not be sent to the landfill.

4.3 International waste management tendencies

The global waste market made an extreme advancement over the last ten years, where the introduction of aspects related to climate protection and conservation of resources has boosted public policies. Politics started to abandon its focus on sanitary landfills and demonstrated that the discussion about recovery of waste can be a tool not only for the mitigation of emissions but also as an industry of generation of secondary resources. In this context, it is possible to observe continues market adaptation followed by integrating new practices and optimizing old ones, as indicated in the following:

- Lack of resources (increase of revenues for secondary resources),
- Increase of selective collection (reduction of calorific value due to the increase of recycling, increase of cardboard per year 5-7% through on line shopping will influence the selective collection based on more volume and decrease of print paper),
- Recycling of material (ban of plastic articles, e.g., shopping bags in Italy; increase of PET per year 3-4% and consequent reduction of glass. Decrease of metal packages 1-3% /yr),
- Decline of conventional incineration in favor of RDF utilization; in 20 years 70% reduction of traditional incineration activities (substitution of 16 million tons of capacity with English and other countries' waste),
- Decline of composting in favor of anaerobic digestion, basically dry technologies,
- Decline of MBT before landfilling in favor to RDF and fuels,
- Prohibition of landfilling of untreated waste,
- Remediation of landfills through the valorization of landfilled waste,
- Biowaste being used as biomass, prohibition to incinerate sludge with other biomass (as coal) in order to not loose phosphate and to reduce organic emissions,
- Valorization of green waste no longer for composting, but for energy recovery at small decentralized power plants and boilers.

5 Conclusions

The National Policy for Solid Waste, issued 2010 in Brazil, raised the topic solid waste management to another level, extrapolating discussions focused exclusively on forms of final disposal on landfills. The new legal framework incorporates the consciousness of wealth and potential possibilities in waste management, and also reveals the errors and omissions that accumulated over the past 30 years. Waste management has changed significantly in recent years, becoming the icon of sustainable development, greatly contributing to environmental protection and, through recycling of waste, also ensuring better environmental conditions.

The new project IKI will provide comprehensive knowledge about the new market in Brazil and construct an interrelationship with the waste sector in Brazil-Germany, establishing an exchange with iconic German institutions on best practices to ensure climate protection and the preservation of natural resources, thus providing a continuous exchange of experiences through vocational and technological education. The support and the dissemination of theory and practical knowledge of German management will bring to Brazil an innovative vision and inspiration to transform the current system in an efficient and continuous reality that meets the premises of the National Solid Waste Policy and the global trends for climate protection. These efforts will cause a cultural change that will protect natural resources and the climate, ensuring a better future for next generations.

6 Acknowledgements

The authors would like to thank the Federal Ministry for the Environment, Nature Conservation, Building and Nuclear Safety (BMUB) for granting the IKI project, and the German Ministry for Economic Cooperation and Development (BMZ) through the German Academic Exchange Service (DAAD) for supporting their participation at the Regional Workshop in Recife, Brazil, September 2016.

7 References

[1] BRASIL. Lei nº 12.305, de 02 de agosto de 2010. Institui a Política Nacional de Resíduos Sólidos; altera a Lei nº 9.605, de 12 de fevereiro de 1998; e dá outras providências. www.planalto.gov.br/ccivil_03/_ato2007-2010/2010/lei/l12305.htm (Accessed on 31/05/2015).

[2] Öko-Institut and IFEU: Climate protection potential in the waste management sector, Study Federal Environment Agency, 61/2010 Dessau (GER), 2010.

[3] Federal Environment Agency (UBA): Submission under the United Nations Framework Convention on Climate Change and the Kyoto Protocol 2016 National Inventory Report for the German Greenhouse Gas Inventory 1990 – 2014, Federal Environment Agency, Dessau (GER) 2016

[4] ABRELPE. Panorama dos resíduos sólidos no Brasil 2015. Abrelpe, São Paulo, 2016.